D1288334

Fungal Conservation
Issues and Solutions

Threats to fungi and fungal diversity throughout the world have prompted debates about whether and how fungi can be conserved. Should it be the site, or the habitat, or the host that is conserved? All of these issues are addressed in this volume, but coverage goes beyond mere debate with constructive guidance for management of nature in ways beneficial to fungi. Different parts of the world experience different problems and a range of examples are presented: from Finland in the North to Kenya in the South, and from Washington State, USA, in the West to Fujian Province, China, in the East. Equally wide-ranging solutions are put forward, from voluntary agreements, through land management techniques, to primary legislation. Taken together, these provide useful suggestions about how fungi can be included in conservation projects in a range of circumstances.

DAVID MOORE is Reader in Genetics in the School of Biological Sciences at the University of Manchester.
MARIJKE NAUTA is a mycologist at the National Herbarium of The Netherlands.
SHELLEY EVANS is a freelance consultant mycologist in the spheres of education, survey and conservation.
MAURICE ROTHEROE is Consultant Mycologist to the National Botanic Garden of Wales and Director of the Cambrian Institute of Mycology.

Fungal Conservation
Issues and Solutions

A SPECIAL VOLUME OF THE BRITISH MYCOLOGICAL SOCIETY

EDITED BY

DAVID MOORE, MARIJKE M. NAUTA,
SHELLEY E. EVANS AND MAURICE ROTHEROE

Published for the British Mycological Society

PUBLISHED BY THE PRESS SYNDICATE OF THE UNIVERSITY OF CAMBRIDGE
The Pitt Building, Trumpington Street, Cambridge, United Kingdom

CAMBRIDGE UNIVERSITY PRESS
The Edinburgh Building, Cambridge CB2 2RU, UK
40 West 20th Street, New York NY 10011-4211, USA
10 Stamford Road, Oakleigh, VIC 3166, Australia
Ruiz de Alarcón 13, 28014 Madrid, Spain
Dock House, The Waterfront, Cape Town 8001, South Africa

http://www.cambridge.org

First published 2001

Printed in the United Kingdom at the University Press, Cambridge

Typeface Times 10/13pt *System* Poltype® [VN]

A catalogue record for this book is available from the British Library

Library of Congress Cataloguing in Publication data

British Mycological Society. Symposium (1999: Royal Botanic Gardens, Kew)
Fungal conservation: issues and solutions: special volume of the British
Mycological Society/edited by David Moore . . . [et al.].
p. cm.
Includes bibliographical references (p.).
ISBN 0 521 80363 2 (hardback)
1. Fungi conservation – Congresses. I. Moore, D. (David), 1942– II. Title.
QK604.2.C66 B75 1999
333.95′2–dc21 00-065169

ISBN 0 521 80363 2 hardback

Contents

		page
List of contributors		vii
Preface		ix
1	Fungal conservation issues: recognising the problem, finding solutions *David Moore, Marijke M. Nauta, Shelley E. Evans & Maurice Rotheroe*	1
2	Current trends and perspectives for the global conservation of fungi *Régis Courtecuisse*	7
3	Conservation and management of forest fungi in the Pacific Northwestern United States: an integrated ecosystem approach *Randy Molina, David Pilz, Jane Smith, Susie Dunham, Tina Dreisbach, Thomas O'Dell & Michael Castellano*	19
4	The future of fungi in Europe: threats, conservation and management *Eef Arnolds*	64
5	Fungi as indicators of primeval and old-growth forests deserving protection *Erast Parmasto*	81
6	Recognising and managing mycologically valuable sites in The Netherlands *Leo M. Jalink & Marijke M. Nauta*	89
7	Threats to hypogeous fungi *Maria Ławrynowicz*	95
8	Wild mushrooms and rural economies *David Arora*	105

9 Threats to biodiversity caused by traditional mushroom cultivation technology in China 111
Siu Wai Chiu & David Moore

10 A preliminary survey of waxcap grassland indicator species in South Wales 120
Maurice Rotheroe

11 Grasslands in the coastal dunes: the effect of nature management on the mycota 136
Marijke M. Nauta & Leo M. Jalink

12 The conservation of fungi on reserves managed by the Royal Society for the Protection of Birds (RSPB) 144
Martin Allison

13 Strategies for conservation of fungi in the Madonie Park, North Sicily 156
Giuseppe Venturella & Salvatore La Rocca

14 Fungal conservation in Ukraine 162
D. W. Minter

15 The threatened and near-threatened Aphyllophorales of Finland 177
H. Kotiranta

16 Fungal conservation in Cuba 182
D. W. Minter

17 Microfungus diversity and the conservation agenda in Kenya 197
P. F. Cannon, R. K. Mibey & G. M. Siboe

18 Fungi and the UK Biodiversity Action Plan: the process explained 209
L. V. Fleming

19 The Scottish Wild Mushroom Forum 219
Alison Dyke

20 The contribution of national mycological societies: establishing a British Mycological Society policy 223
David Moore

21 The contribution of national mycological societies: the Dutch Mycological Society and its Committee for Fungi and Nature Conservation 242
Marijke M. Nauta & Leo M. Jalink

22 Fungal conservation in the 21st century: optimism and pessimism for the future 247
David Moore, Marijke M. Nauta, Shelley E. Evans & Maurice Rotheroe

Index 256

Contributors

Martin Allison
The Royal Society For The Protection Of Birds, The Lodge, Sandy, Bedfordshire SG19 2DL, UK

Eef Arnolds
Holthe 21, 9411 TN Beilen, The Netherlands
<eefarnolds@hetnet.nl>

David Arora
343 Pachecho Avenue, Santa Cruz, Santa Cruz, California 95062, USA
<arora_david@yahoo.com>

P. F. Cannon
CABI Bioscience, Bakeham Lane, Egham, Surrey TW20 9TY, UK
<p.cannon@cabi.org>

Michael Castellano
USDA Forest Service, Pacific Northwest Research Station, Forestry Sciences Laboratory, 3200 SW Jefferson Way, Corvallis, OR 97331, USA

Siu Wai Chiu
Department of Biology, Chinese University of Hong Kong, Hong Kong S.A.R., China
<siuwaichiu@cuhk.edu.hk>

Régis Courtecuisse
Département de Botanique, Faculté des Sciences Pharmaceutiques et Biologiques, B.P. 83, F-59006 Lille Cedex, France
<rcourtec@phare.univ-lille2.fr>

Tina Dreisbach
USDA Forest Service, Pacific Northwest Research Station, Forestry Sciences Laboratory, 3200 SW Jefferson Way, Corvallis, OR 97331, USA

Susie Dunham
Department of Forest Science, Oregon State University, Corvallis, OR 97331, USA

Alison Dyke
Woodland Ecology and Non Timber Forest Products, 3F1 46 Broughton Road, Edinburgh EH7 4EE
<alison.dyke@talk21.com>

Shelley E. Evans
Icknield House, 8 Saxonhurst, Downton, Salisbury, Wilts SP5 3JN
<Shelley-Evans@myco-services.freeserve.co.uk>

L. V. Fleming
Joint Nature Conservation Committee, Monkstone House, City Road, Peterborough PE1 1JY, UK
<Flemin_v@jncc.gov.uk>

Leo M. Jalink
National Herbarium of The Netherlands, University Leiden Branch, P.O. Box 9514, 2300 RA Leiden, The Netherlands

H. Kotiranta
Finnish Environment Institute, P.O.Box 140, FIN-00251 Finland
<heikki.kotiranta@vyh.fi>

Salvatore La Rocca
Dipartimento di Scienze Botaniche, Università di Palermo, Via Archirafi 38, I-90123 Palermo, Italy

Maria Ławrynowicz
Department of Algology and Mycology, University of Łódź, Banacha 12/16, PL-90–237 Łódź, Poland
<miklaw@biol.uni.lodz.pl>

R. K. Mibey
Department of Botany, University of Nairobi, P.O. Box 30197, Nairobi, Kenya

D. W. Minter
CABI Bioscience, Bakeham Lane, Egham, Surrey, TW20 9TY, UK
<d.minter@cabi.org>

Randy Molina
USDA Forest Service, Pacific Northwest Research Station, Forestry Sciences Laboratory, 3200 SW Jefferson Way, Corvallis, OR 97331, USA
<rmolina@fs.fed.us>

David Moore
School of Biological Sciences, 1.800 Stopford Building, The University of Manchester, UK
<david.moore@man.ac.uk>

Marijke M. Nauta
National Herbarium of The Netherlands, University Leiden Branch, P.O. Box 9514, 2300 RA Leiden, The Netherlands
<Nauta@nhn.leidenuniv.nl>

Thomas O'Dell
USDA Forest Service, Pacific Northwest Research Station, Forestry Sciences Laboratory, 3200 SW Jefferson Way, Corvallis, OR 97331, USA

Erast Parmasto
Institute of Zoology and Botany, Estonian Agricultural University, 181 Riia Street, 51014 Tartu, Estonia
<e.parmasto@zbi.ee>

David Pilz
USDA Forest Service, Pacific Northwest Research Station, Forestry Sciences Laboratory, 3200 SW Jefferson Way, Corvallis, OR 97331, USA

Maurice Rotheroe
Fern Cottage, Falcondale, Lampeter, Dyfed SA48 7RX, UK
<m.rotheroe@btclick.com>

G. M. Siboe
Department of Botany, University of Nairobi, P.O. Box 30197, Nairobi, Kenya

Jane Smith
USDA Forest Service, Pacific Northwest Research Station, Forestry Sciences Laboratory, 3200 SW Jefferson Way, Corvallis, OR 97331, USA

Giuseppe Venturella
Dipartimento di Scienze Botaniche, Università di Palermo, Via Archirafi 38, I-90123 Palermo, Italy
<gvent@unipa.it>

Preface

Conservation is a major concern at the moment. In the second half of the twentieth century, naturalists (including mycologists) became aware of a general decline in natural habitats. Significant decline in the number of species, and in the occurrence of species of fungi, was detected in the 1960s in several countries in Europe and was correlated with changes in land usage and with three consistent themes of environmental pollution: eutrophication (contamination of water bodies with nutrients draining off agricultural land), acid rain (mostly downwind of industrial smoke plumes) and desertification (caused by shifts in precipitation patterns resulting from climatic change). Since that time the general attitude has changed to one in which managers try to improve biodiversity.

This book examines the various problems associated with fungal conservation. In different parts of the world there are several threats to fungi and fungal diversity that prompt thoughts of conservation. However, it is not self-evident whether and how fungi themselves can be conserved. Perhaps the emphasis should be placed on conservation of the site, or the habitat, or the host? All of these issues are addressed and debated here, but this book goes beyond mere debate by filling the need for practical guidance for management of nature in ways beneficial to fungi. Different parts of the world experience different problems, so there are different solutions too. Here we present a range of different case studies and describe the range of regulatory and control procedures that are beginning to emerge. Taken together, these give tips on how to include fungi in conservation projects in a range of circumstances.

We believe that the combination of reviews of the research literature with discussion of how to solve problems in the real world makes this book virtually unique, at least in English. We hope, therefore, that it will be found to be a succinct and practical guide about the subject – one that will be used just as much by nature managers, administrators and politicians who have to make policy to deal with conservation issues, as by amateur and professional mycologists who want to know more about the status of the fungi.

The book has its origin in a British Mycological Society Symposium which was organised by Bruce Ing, Maurice Rotheroe and Shelley Evans at the Royal Botanic Gardens in Kew on 13 November 1999, entitled *Fungal Conservation in the 21st Century.* These contributions have been integrated with a selection of the most appropriate papers delivered at the XIIIth Congress of European Mycologists, held in Alcala de Henares, Spain, 21–25 September 1999. Importantly, the conference contributions have been complemented by six specially commissioned chapters. Among these are chapters that enhance the geographic representation of the book's contents so that it is truly global in its coverage. We have also added two chapters that relate how national mycological societies can contribute to fungal conservation and two editorial overviews. Chapter 1 attempts to provide some signposts for readers to help them get the best out of the rest of the book. The final chapter is an attempt to convey in writing some feeling of the discussions that occurred during the meetings. We asked contributors to respond to the question 'Are you optimistic or pessimistic about fungal conservation in the 21st century?' They reply from different standpoints, but, thankfully, they all count themselves as optimists. So, read on and find out why!

David Moore
Marijke Nauta
Shelley Evans
Maurice Rotheroe

October 2000

1

Fungal conservation issues: recognising the problem, finding solutions

DAVID MOORE, MARIJKE M. NAUTA, SHELLEY E. EVANS & MAURICE ROTHEROE

Nonbiologists may be excused for questioning whether microbial diversity is really under threat. At a superficial level, micro-organisms seem to be tolerant of almost any set of conditions thrown at them. Also, they appear to have reproductive capacities able to generate populations of truly astronomic numbers in very little time. However, that *is* a superficial understanding and any belief that microbial species are not threatened is simply wrong.

James T. Staley of the University of Washington gave his answer to the challenge 'Microbiologists are not concerned with endangered species, are we?' in a commentary published in 1997 (Staley, 1997). His simple answer to this question is 'Yes, some microbial species are threatened', but the argument Staley develops is interesting and has some valuable points for mycologists to ponder. Even though the commentary was written largely from the bacteriologist's point of view, Staley mentions lichens and fungi so it is clear that he does include even the mushrooms and toadstools within his definition of micro-organisms. This is useful for us as we attempt in this brief introductory chapter to highlight and provide cross-references to the wide variety of aspects of fungal conservation that are included in this book. We are not alone in the belief that such topics are important!

Indeed, Staley puts the level of importance very high. Micro-organisms produced the original biosphere of Earth. 'Not only have they made conditions suitable for the evolution and existence of macroscopic life forms, but they also continue to drive and profoundly influence many of the essential biogeochemical cycles' (Staley, 1997). Furthermore, most of the present-day biodiversity among the eukaryotes is microbial, being generated by the protists, algae and fungi. Bacteria, of various sorts, provide the biodiversity within the prokaryotes, of course. So the consequence is that 'the tree of life is largely a tree of microorganisms . . . much of the diversity on Earth is microbial with the plants and animals appearing as small, terminal branches' (Staley, 1997).

This fundamental importance of micro-organisms can hardly be doubted, so why is there so little general interest in conservation of microbes? Staley (1997) puts this down to something he calls 'kinship', claiming that humans share strong kinship with many animals and plants, a kinship which can blossom into fondness for closely related and 'warm and fuzzy' animals. Microbes, though, are generally too small to be noticed much by humans, even though human lives are daily more closely intertwined with microbes than with any other organisms. Further, microbes evoke negative feelings because they are associated with disease and spoilage. Finally, there is a general ignorance about the degree to which our daily lives depend on the beneficial activities of many microbes – from sewage sludge through to agriculture, and the making of bread, and antibiotics and other life-saving drugs. 'Because microorganisms rank so low on the kinship scale, the demise of a microbial species is not an emotional issue for humans' (Staley, 1997). We do not expect many to rally to a cry to 'Save the whale's intestinal microbes'!

Staley (1997) suggests that the general phenomenon is that a micro-organism is threatened when its ecological niche is threatened. Consequently, 'the most satisfactory manner in which to preserve the organisms is through protection of the environment and thereby the natural community itself' (Staley, 1997). However, Staley acknowledges that 'we have described so few species; many species may be threatened whose existence are still unknown.' And his final conclusion is that 'Our knowledge of microbial diversity . . . is so meagre that we do not yet know if and when most species are threatened. . . . Our very inability to answer the question of threatened microbial species cries loudly for the need for microbial systematists and ecologists to begin to address the exciting challenges regarding our knowledge of the extent of microbial diversity on Earth' (Staley, 1997).

That brings us to the first point we wish to highlight from this book. It is most succinctly stated by David Minter in Chapter 16 (p. 193): 'In many parts of the world mycologists are an endangered species. It follows that fungal conservation can only occur if mycologists are conserved.' But other authors express similar opinions. Régis Courtecuisse (p. 10) puts it this way: 'Incidental problems and questions around inventories which have to be considered are (a) promoting the conservation of taxonomists themselves . . . ' and Eef Arnolds (p. 77) like this: 'It is obvious that conservation of fungi depends on the input of mycologists. But at present it seems to be also the other way around: the future of mycologists depends on their input in conservation.' The root problem is that our level of ignorance is so great that we do not have the numbers of experts needed to make serious contributions to knowledge of species sufficiently quickly to conserve those species. 'Taxonomists are scarce because of a shift in academic programmes toward molecular systematics and ecology' (Randy Molina *et al.*, p. 39). One might also add that for several years now funding agencies around the world have been operating a similarly skewed funding policy. Mycological research is rarely funded, anyway, because a lower value judgement is placed upon it than is applied to similar research on lower animals or lower plants. Another aspect, perhaps, of the lack of kinship to which reference is made above. In the long term these attitudes must change and the importance of the kingdom of fungi recognised sufficiently to assure equitable funding for its study. To a very large extent this is a matter of public education and several of our authors mention this. Régis Courtecuisse mentions the need for public education (Chapter 2, p. 14), and David Moore and Siu Wai Chiu claim that 'Education is the key' in China (Chapter 9, p. 118).

It will take a long time for an education policy to result in significantly more experts with attitudes changed sufficiently for the value of fungal biology to be fully appreciated. In the meantime we have the real world to deal with – a real world in which those mycologists who do exist may be prevented from making a full contribution by poor infrastructure or political and economic isolation. David Minter, in Chapter 14 (p. 164), illustrates how effective voluntary help (in this case through provision of second-hand computers) together with intergovernmental assistance (through the UK Government's Darwin Initiative Programme) is enabling Ukrainian mycologists to complete the databases and surveys that are essential to effective national conservation policies. He tells a similar story in Chapter 16 (p. 192), although in this case Cuban mycologists are suffering the inevitable shortages and isolation resulting from a unilateral economic blockage imposed by the USA. Again, provision of resources (another Darwin Initiative Project) enables local mycologists to progress towards a national fungal conservation strategy.

For more immediate input, particularly to projects under way now, 'The depleted ranks of classical fungal taxonomists can be augmented, however, by a cadre of experienced parataxonomists, people with less formal schooling in mycology, who are trained and gain significant experience in fungal identification' (Molina *et al.*, Chapter 3, p. 39). Similar ideas, perhaps, emerge from the Dutch experience in raising interest which Leo Jalink and

Marijke Nauta suggest makes it evident that 'managers need clear instructions' about mycologically valuable sites (Chapter 6, p. 90).

If the information is provided, landowners, managers and administrators have considerable sympathy for including fungi in their conservation management. Indeed, it seems a sensible strategy for mycologists to be proactive in establishing collaborations with those involved in land management and, especially, with groups concerned with conservation of other organisms (see Martin Allison, Chapter 12, p. 144). There is certainly no excuse for mycologists being short of cogent arguments for inclusion of fungi in conservation schemes. Randy Molina *et al.* (Chapter 3, p. 23) detail the four themes that need to be emphasised when 'educating land managers . . . is vital'. Additional ready-to-use material can be found in Chapter 17, in which Cannon *et al.* discuss, largely from the point of view of population ecology, 'Why are fungi difficult to conserve' (p. 198) and 'Why are fungi important' (p. 199).

We know very little about fungal population biology; in fact, even less about fungal population genetics. Randy Molina *et al.* (Chapter 3, p. 25) discuss the role of fungi in communities and describe projects aimed at determining the population genetics of representative species (p. 33). Related to this is the detailed analysis of the population biology of *Lentinula edodes* that shows how the traditional cultivation method in China (especially outdoor cultivation accompanied by harvesting at maturity) is likely to endanger both the cultivars and the wider gene pool of the wild mushroom (David Moore and Siu Wai Chiu, Chapter 9, p. 113).

The main tools available to the fungal conservationist are outlined first by Régis Courtecuisse (Chapter 2, p. 10) to be inventories (checklists), mapping programmes, and Red Data lists. These being the crucial aspects of fungal conservation, they appear in some guise in all chapters. Particularly helpful discussions can be found in Chapters 3 (p. 35), 4 (p. 70), 5 (p. 83), 6 (p. 90), and 17 (p. 202). Eef Arnolds (Chapter 4, p. 66) also discusses the species concept – an important issue for any survey, whilst Molina *et al.* (Chapter 3, p. 43) describe 'habitat modelling' as a tool in conserving fungal resources. Examples of survey work are given in Chapters 3 (p. 19), 5 (p. 81), 6 (p. 89), 7 (p. 95), 9 (p. 111), 10 (p. 120), 11 (p. 136), 13 (p. 156), 15 (p. 177), 16 (p. 182), and 17 (p. 197).

Surveys and mapping programmes culminate in the production of Red Data lists. Although 'Red Data' in this phrase usually carries with it the danger connotation commonly linked with the colour red, it's important to remember that in this case the word is an acronym, the full phrase being Rarity, Endangerment and Distribution Data lists. This *is* important because the full phrase shows explicitly the amount of information which is required to make the judgement about whether or not to include a species in a Red Data list. Red Data lists are discussed to some extent in most chapters, especially those already highlighted as dealing with surveys. However, Maria Ławrynowicz (Chapter 7, p. 96) shows how different national Red Data lists can be integrated to reach wider conclusions, while Giuseppe Venturella and Salvatore La Rocca (Chapter 13, p. 156), and Heikki Kotiranta (Chapter 15, p. 177) illustrate how local surveys can be compared, on the one hand with an international Red Data listing, and on the other hand with international Red list categories.

Conservation strategies emerge at a variety of levels and provide examples which might be applicable elsewhere. Molina *et al.* (Chapter 3, p. 20) outline the US Federal laws regulating forest management, mentioning the different goals of the different agencies involved. A different set of conflicts (and their resolution) discussed by Martin Allison (Chapter 12, p. 153) is that which can arise 'within conservation management when one group of animals or plants is favoured above another.' Vincent Fleming (Chapter 18, p. 209) details the UK response to the *Convention on Biological Diversity* – essentially the administrative mechanics of conservation in the UK. Below the governmental level, David Moore (Chapter 20, p. 223), and Marijke Nauta and Leo Jalink (Chapter 21, p. 242) show how two national mycological societies (the British and Dutch mycological societies respectively) have reacted and developed programmes aimed at conserving fungi. In Chapter 19 (p. 219) Alison Dyke reports how a purely voluntary code of practice has been established directly by the groups involved in wild mushroom harvesting in Scotland. A range of wild harvested fungal fruit bodies command prices that make them worth shipping over intercontinental distances, as discussed by David Arora (Chapter 8, p. 105), so this code of practice may be applicable elsewhere. In contrast, the commercial harvest of edible forest mushrooms is controlled by Federal laws in the United States (Molina *et al.*, p. 46, and see Eef Arnolds, p. 76).

The Scottish Mushroom Forum's code of practice (Table 19.3, p. 221) is one of several examples of specific advice and instruction included in this book. Others are a 'set of summary statements' for use when 'planning and conducting conservation efforts for fungi' (Molina *et al.*, Chapter 3, p. 54); some management guidelines from Leo Jalink and Marijke Nauta (Table 6.2 and Table 6.3, p. 93); and British Mycological Society codes of practice (Chapter 20, p. 235).

With these, and other, explicit pieces of advice based upon practical experience, we hope that this book will make a *constructive* contribution to fungal conservation. It is a global problem and we include examples from Finland in the North to Kenya in the South, and from Washington State, USA, in the West to Fujian Province, China, in the East. Our authors identify threats faced by fungi of all types. Inevitably, even though 'It is probably true to say that the majority of fungi would be describable as "microfungi"' (Paul Cannon *et al.*, Chapter 17, p. 197) descriptions of work with larger fungi – truffles and mushrooms – tend to predominate. The balance, of course, is governed by the research which is being done and the research interests of those doing it.

Our authors also suggest solutions ranging from voluntary agreements, through 'fungus-favourable' land management practices, and on to primary legislation. We have to stress that this book cannot give ready-made solutions to all the problems that might arise concerning conservation of fungi. What we have assembled is a set of descriptions of how far we have got with conservation of fungi, with some focus on the bottlenecks that remain, and with a range of guidelines that may help in improving conservation of fungi in the future. The bottom line, though, is quite clearly that 'Conservation of fungi is, like conservation of other organisms, in the very first place conservation of their habitats combined with adequate management' (Eef Arnolds, Chapter 4, p. 72). Save the world and we'll save the fungi with it. Conserve the fungi and your one and only planetary home will be equally safe.

Reference

Staley, J. T. (1997). Biodiversity: are microbial species threatened? *Current Opinion in Biotechnology* **8**, 340–345.

2

Current trends and perspectives for the global conservation of fungi

RÉGIS COURTECUISSE

Introduction

For rather more than a year, the specialist group for fungi within the Species Survival Commission (SSC) of the International Union for the Conservation of Nature (IUCN) has been revived after interruption in its activities between 1995 and 1998. As a member of the European Council for the Conservation of Fungi (ECCF) standing committee, I shared with my ECCF colleagues the regret that fungi were no longer taken into account within the IUCN. So we decided to take advantage of the June 1998 *Planta Europa* meeting, in Uppsala, to establish new contacts with the IUCN and to revive this specialist group.

Within the new specialist group for fungi, the ECCF serves as the main framework since it has accumulated much data and experience on the topic since 1984. But I am also trying to federate further mycologists involved in conservation outside Europe so that the group will consist of a genuinely enlarged and international network.

In this chapter I will give a brief description of the present global state of knowledge concerning fungal conservation and indicate the main priorities we should consider for the future. Of course, this owes a great deal to the ECCF heritage, especially through the decisive contributions from some of its members, such as Eef Arnolds and others. I also received feedback from some members of the new group but I do not claim that the literature survey is in any way exhaustive and I do realise that some points may not be represented here. The specialist group of which I serve as chairman is currently still 'under construction' and we attach great importance to communication between all people who are interested or involved in fungal conservation and depend on them for further information.

Current status and problems

Conservation status of fungi

Fungi are very seldom legally protected. Such a situation does exist in Slovakia, where 52 species have a 'special legal status', enabling managers to prevent damage to their habitats (Lizon, 1999). Where no legal status is available, some voluntary efforts have been made to produce codes of practice or have advisory documents published, stressing the importance of fungal conservation and summarising the recommendations to achieve that. This has been done, for example, in Switzerland (Egli *et al.*, 1995), and in the United Kingdom (English Nature, 1998; and see Chapters 18 to 21).

Nevertheless, even without a legal conservation status, Red Data lists have been published, or are currently in preparation, in many countries, especially in Europe. This demonstrates the increasing concern of mycologists for this topic. At the same time, more and more nature managers pay increasingly greater attention to fungi.

So, the idea that fungi may be threatened and deserve special attention to their conservation is now well established, at least in Europe, where books and symposia have been devoted to the topic in relation to environmental problems (Frankland, Magan & Gadd, 1996; Rotheroe, 1996*a*). But this is not yet true everywhere.

Current problems

The main threats and causes of decline in fungi world-wide can be categorised as resulting either from global, or specific or local problems. Air pollution, considered globally, may influence fungi through the greenhouse effect producing a slow climate change, as well as exerting an indirect effect through modification of vegetation. Such changes might threaten climate-sensitive species and/or favour the development of more thermophilic taxa which could in turn act as alien competitors against native species. The balance between parasites and their hosts might also be changed (Lonsdale & Gibbs, 1996; Pettitt & Parry, 1996). Some northward migrations have been demonstrated in recent years, especially in the western part of Europe. Elevation of the sea level could also become a problem for species inhabiting coastal ecosystems (Rotheroe, 1996*b*). Another global problem widely recognised to be of paramount importance in the decline of fungi world-wide is the destruction of habitats and the dramatic felling of forests, particularly in the tropics, but also representing a potential problem in more temperate or even sub-boreal to boreal forests.

Some problems with a more local or specific effect on fungi include: air pollution, which, whatever the scale, causes well-documented species decline; deposition of various pollutants, leading to soil modifications (the mechanism of which is not clearly understood); accumulation of metals and other pollutants interfering with fungi, either macrofungi (Ing, 1996) or microfungi (Boddy *et al.*, 1996; Magan, Smith & Kirkwood, 1996). Fragmentation of habitats is also a major problem. It has been shown (e.g. Chaumeton, 1994) that fragmentation of habitats makes it difficult for some species to maintain normal population. Such a situation often results from forest cutting, urban extension, evolution of land uses and change of agricultural practices. In particular, modern agriculture increasingly uses chemical treatments that can give rise to various kinds of environmental modifications. This problem is especially well documented in Europe for grassland fungi, both macromycetes such as *Hygrocybe* spp. (Rotheroe *et al.*, 1996) but also soil fungi (Bardgett, 1996). A more specific problem concerns edible species. When harvesting is done by individuals or families, it seldom leads to concern for survival of the species (Egli, Ayer & Chatelain, 1990). On the other hand, the commercial harvesting of the same species causes great concern, as is evident from the numerous publications on the subject, especially in North and Central America (Pilz & Molina, 1996, 1997; Palm & Chapela, 1997; Rowe, 1997). This is a controversial topic. Some people claim that heavy harvesting is harmless (Arora, 1999*a*, 1999*b*) and others claim the opposite (Rotheroe, 1998), depending mainly on the harvesting tools, but also taking into account the longer term, for which we have insufficient background knowledge to know what will really happen in the future. Furthermore, there is currently little scientific evidence about the impact of harvesting on the mycelium itself and more studies are needed on that topic. Suggestions for voluntary and regulatory control have been published also in Europe, such as Switzerland (Keller, 1991), the United Kingdom (Leonard, 1997*a*,*b*) or some parts of France. ECCF members from eastern Europe complain about the ecological damage caused to their forests by commercial harvesting of edible fungi that are exported to western Europe (see, for example, Ivancevic, 1998; Pop, 1998). Further problems arise from the harvesting of edible chanterelles (in part undescribed species) from Africa, tonnes of which are imported each year into France, for example. A similar problem may be faced by the decorative tropical polypores that are used in floral compositions. We need to know more about the actual places and conditions of harvesting of these species.

What can we do?

Looking at these causes of fungal decline, we might feel discouraged because, as fungal conservationists, we have no opportunity to manage global warming or global forest felling. Nor are we likely to have much influence on the politics that affect agricultural practices or industrial pollution. It is not even clear how we should handle the harvesting of edible fungi because important economic interests are involved – commercial interests, of course, but sometimes also at the individual family level, thus involving a social dimension in the problem. What we can actually do, at our 'mycological level', concerns fungi in the field, and is related to our knowledge about the ecological and heritage value of fungi. The 'heritage' dimension of fungi in this context refers to their potential for indicating the holistic natural value of a given site or habitat in a way that integrates many inputs including historical, nature management, biodiversity, and conservation aspects. It is our task to popularise the value of this heritage quality of fungi and to use it as a force in nature conservation. For that, we have some useful tools at hand.

Our tools

First are inventories. In most of the countries of the world, what are basically lacking are checklists of fungi. This is not only true of tropical or developing countries, but also of the first world's great nations. The situation is even worse at the continental level. A project to initiate a European checklist has been urged by the IUCN and is currently being considered by the ECCF for funding in the near future. The first priority in our tasks as fungal conservationists is to promote fungal inventories. Incidental problems and questions around inventories which have to be considered are (a) promoting the conservation of taxonomists themselves, (b) developing arguments in favour of such inventories, (c) using a reasonable consensus taxonomy which is discerning enough to maintain the ecological bioindicative value of the taxa recognised.

Second are mapping programmes. Inventories provide an idea of the fungal diversity: mapping programmes should yield useful information about the rarity of species, their eventual decline and finally their heritage value. Mapping is the second important task we must promote as fungal conservationists. Inventories and mapping programmes are intimately connected, of course. Mapping programmes are conducted in many countries where inventory data are sufficient, as well as some in which inventory data are not sufficient, for example in Australia where an interesting and ambitious project is already running (<http://calcite.apana.org.au/fungimap/>). In Europe the ECCF is currently organising a mapping programme of endangered species (Otto & Ohenoja, 1998).

Third are Red Data lists. On the basis of inventory and mapping programmes, mycologists are able to prepare Red Data lists for various geographical or political units. This is the third fundamental task for fungal conservationists. Red Data lists have appeared mainly in Europe, at the continental as well as national level (Ing, 1993*a*,*b*) but also recently in the USA, for example, with a preliminary list for Oregon (Castellano, 1997). Red Data listing raises many problems, especially the crucial choice of criteria for putting species on the list or not. Several models are in current use but there is a strong desire to use a homogeneous system of criteria, at least at the national level where comparability with neighbouring countries is a priority. We prefer to use the IUCN criteria although the last edition of these (1994) was clearly made for big African mammals such as elephants or rhinoceros. Criticism of these criteria has been quite strong among mycologists and some other naturalists, and attempts are currently being made to revise and adapt them to fungal characteristics (Arnolds, 1998; Kotiranta, 1998, 1999). Some interesting ideas promoting the use of unusual but useful criteria are also in current development (Courtecuisse, 1997; and see Chapter 18). A special commission has been established within the ECCF for that purpose, the work of which is currently ongoing.

Conservation strategies

The results of such programmes, even if still quite fragmentary, enable us to consider fungal conservation and to include fungi in nature conservation. An important point is that, as widely admitted, fungi cannot be protected individually, as can sometimes be done with plants or animals, by publishing lists of legally protected species. The biology of fungi, the frequent difficulty that even trained people have in identifying them, and other peculiarities, do not fit with this option.

In situ *conservation by conserving natural habitats*

The most important tool for fungal conservation is the conservation of their habitats. This was cited as the first priority by all the contacts I consulted while preparing this chapter, from Europe and from overseas and tropical countries as well.

So the first strategy we must develop is to promote the conservation of natural habitats, especially in view of the current tendency to replace native forests by alien trees that are more valuable timber producers, a problem also in some tropical countries. More locally, some important places may be in need of protection, partly on a mycological basis (see for example Rotheroe, 1995).

In proposing a hierarchy of habitats for the purpose of fungal conservation, we must consider whether or not we have enough mycological data. If such data are available, interesting methods currently being developed can be used to 'weight' species according to their heritage value, inferred from their status in Red Data lists for example. Such an approach has been developed in The Netherlands (see Chapter 7) and also in France (Courtecuisse & Ansart-Chopin, 1997; Courtecuisse & Blot-Quénu, 1998). Where sufficient data are unavailable, a provisional position is to concentrate on habitats that are interesting and rich for other taxonomic groups, plants for example, with which fungi are frequently associated. If the congruence of vegetation communities with communities of different types of fungi is high enough, conserving the plant community will 'take along' all the associated fungi, even including the undescribed species. Other methods for using incomplete inventories are being developed, especially in France.

In situ *conservation on mycological reserves*

The concept of mycological reserves is also interesting since it promotes the conservation of habitats which are especially valuable for fungi and which are not already considered for conservation otherwise. Such reserves already exist in some European countries and much work is being done at the moment in France on this concept. Much remains to be done as to the desirable size of such mycological reserves so that they prove to be efficient for fungal populations (see, for example, Senn-Irlet, 1998).

In situ *conservation with ecological corridors*

In relation to problems of habitat fragmentation, the concept of biological corridors should also perhaps be considered for fungi. As far as I know, very few studies have been dedicated to this kind of research in mycology. Only Alexander & Watling (1987) considered the role of *Betula* within Sitka spruce forests. Biological corridors have strongly positive effects on the survival and even on the restoration of populations of various plants

Table 2.1. *Current membership of the SSC/IUCN Specialist Group for Fungi*

Chair	Régis Courtecuisse (France)
Secretary	Claudia Perini (Italy)
Ordinary members	Anders Bohlin (Sweden)
	Peter Buchanan (New Zealand)
	Julietta Carranza (Costa Rica)
	Ignacio Chapela (Mexico)
	Shelley Evans (UK)
	Ana Franco (Columbia)
	Bruce Ing (UK)
	Teresita Iturriaga (Venezuela)
	Heikki Kotiranta (Finland)
	Pavel Lizon (Slovakia)
	Jean Lodge (USA – Porto Rico)
	Tom May (Australia)
	Greg Mueller (USA)
	Marijke Nauta (The Netherlands)
	Lorelei Norvell (USA)
	Scott Redhead (Canada)

and animals. It is quite likely that they may also have a positive effect on fungal populations.

Ex situ *conservation*

In some cases, especially for saprotrophic species growing in culture, *ex situ* conservation programmes could be developed. A good example of this has been shown for *Pleurotus nebrodensis* in Sicily (Venturella, 1999) and more generally by Homolka *et al.* (1999). This possibility might be of great importance for species on the verge of extinction or those strictly confined to very limited or threatened habitats. This concept is rather similar to the one used for plants, through seed banks.

The SSC/IUCN specialist group for fungi

After reviewing the problems and some possibilities for conserving fungi, I would like to add some words on the SSC/IUCN specialist group for fungi, its composition and its role, at least as I understand this.

The composition of this group at the time of writing is shown in Table 2.1. As can be seen, the group lacks members from Africa and Asia. I have had contacts with several colleagues from these continents but, at the moment, no name is readily available. We hope, of course, that the group will be enlarged in the future.

Future tasks

The main roles of this specialist group are to promote ideas about fungal conservation and its various implications and also to promote research in the different possibilities related to the goals and tools I have summarised here. As chairman of this group, I will encourage research in the crucial areas of inventories, mapping programmes and Red Data listing programmes. These topics represent the basic research we have to develop in all countries.

We will also have to develop specific research about the objective definition of interesting and priority habitats for fungi. Further research is also necessary and urgent to develop and promote the use of fungal bioindicators of environmental quality as a tool to be used by nature managers. The research concerning indicator species of old and primeval forests (see Parmasto & Parmasto, 1997; Tortic, 1998; O'Dell, Kiester & Molina, 1999, among other examples) must be continued in this respect. We should also strengthen the promotion of the fungal role, especially mycorrhizal species, in rehabilitation of polluted (e.g. Allen, Espejel & Sigüenza, 1997) or weakened (Janos, 1996) areas. Further fundamental study is needed on special topics like the conservation of parasitic fungi which are also important for the holistic functioning of ecosystems, or the role of natural history collections in fungal and nature conservation (e.g. May, 1998).

It is necessary, too, to establish, strengthen and regularise contacts with other naturalists involved in conservation, in order to exchange data and create synergistic arguments for conservation of globally interesting habitats or places which can be made to administrations and decision holders at any level, local, regional, national and supra-national. For example, at the European level, contacts with the Bern convention structure are currently in progress and will hopefully soon lead to the long-awaited integration of fungi into the Bern Convention appendix. At a different level, we should also reflect on social problems, like the consequences of putting controls on harvesting of edible fungi, for example.

Finally, one of our tasks will be to build an efficient network for developing public education. Communication between the members of our group must be made easier – of course through electronic tools but also through the journal *Fungi and Conservation Newsletter*, three numbers of which were published in the past by the IUCN and which I intend to try to revive.

Conclusion

As a conclusion, I would like to thank and congratulate the friends and colleagues who have conducted or chaired the ECCF so far. The work of these people has provided a strong framework for fungal conservation that remains to be developed fully in Europe, where the idea was born, and to be used as a model elsewhere, with inevitable local adaptations. That is what I will try to do in the course of my position as chair of the ECCF at the European scale on the one hand and of the SSC/IUCN Specialist Group for Fungi on a wider basis on the other. I trust that continued friendly collaboration will yield further positive developments and progress in fungal conservation for the new century.

References

Alexander, I. J. & Watling, R. (1987). Macrofungi of Sitka spruce in Scotland. *Proceedings of the Royal Society of Edinburgh* **93**, 107–115.

Allen, E. B., Espejel, I. & Sigüenza, C. (1997). Rôle of mycorrhizae in restoration of marginal and derelict land and ecosystem sustainability. In *Mycology in Sustainable Development: Expanding Concepts, Vanishing Borders* (ed. M. E. Palm & I. H. Chapela), pp. 147–159. Parkway Publishers: Boone, NC, USA.

Arnolds, E. (1998). A new red list of macrofungi in The Netherlands, based on quantitative criteria. In *Conservation of Fungi in Europe*, Proceedings of the 4th meeting of the European Council for the Conservation of Fungi, Vipiteno (Sterzing), Italy, 9–14 September, 1997 (ed. C. Perini), pp. 17–34. Privately published. University of Sienna.

Arora, D. (1999*a*). The way of the wild mushroom. <http://www.calacademy.org/calwild/fall99/mushroom.htm>

Arora, D. (1999*b*). The global mushroom trade. <http://www.calacademy.org/calwild/fall99/mushroom.htm>

Bardgett, R. D. (1996). Potential effects on the soil mycoflora of changes in the UK agricultural policy for upland grasslands. In *Fungi and Environmental Change* (ed. J. Frankland, N. Magan & G. M. Gadd), pp. 163–183. Cambridge University Press: Cambridge, UK.

Bermingham, S. (1996). Effects of pollutants on aquatic hyphomycetes colonizing leaf material in fresh waters. In *Fungi and Environmental Change* (ed. J. Frankland, N. Magan & G. M. Gadd), pp. 201–216. Cambridge University Press: Cambridge, UK.

Boddy, L., Frankland, J. C., Dursun, S., Newsham, K. K. & Ineson, P. (1996). Effects of dry-deposited SO_2 and sulphite on saprotrophic fungi and decomposition of tree leaf litter. In *Fungi and Environmental Change* (ed. J. Frankland, N. Magan & G. M. Gadd), pp. 70–89. Cambridge University

Press: Cambridge, UK.

Castellano, M. A. (1997). Towards a Red List for Oregon macrofungi. In *Conservation and Management of Native Plants and Fungi* (ed. T. N. Kaye, A. Liston, R. M. Love, D. L. Luoma, R. J. Meinke & M. V. Wilson), pp. 222–226. Native Plant Society of Oregon: Corvallis.

Chaumeton, J. P. (1994). Impact du morcellement du milieu naturel sur la flore fongique. Thèse Docteur des Sciences (Université Paul Sabatier, Toulouse).

Courtecuisse, R. (1997). Liste rouge des champignons menacés de la région Nord, Pas-de-Calais (France). *Cryptogamie, Mycologie* **18**, 183–219.

Courtecuisse, R. & Ansart-Chopin, S. (1997). Les champignons de la forêt de Desvres (Pas-de-Calais). Analyse patrimoniale et conservatoire. *Bulletin de la Société Mycologique du Nord* **62**, 7–24, 41–53.

Courtecuisse, R. & Blot-Quénu, C. (1998). Contribution à l'inventaire mycologique de la réserve biologique de Merlimont (Pas-de-Calais). *Bulletin de la Société Mycologique du Nord* **63**, 5–30.

Egli, S., Ayer, F., Lussi, S., Senn-Irlet, B. & Baumann, P. (1995). La protection des champignons en Suisse. Un aide-mémoire à l'intention des autorités et des milieux concernés. *Notice pour le praticien* 25, 8 pp. Institut Fédéral de recherches sur la forêt, la neige et le paysage.

Egli, S., Ayer, F. & Chatelain, F. (1990). Der Einfluss des Pilzsammelns auf die Pilzflora. *Mycologia Helvetica* **3**, 417–428.

English Nature (1998). The conservation of wild mushrooms. The wild mushroom pickers' Code of Conduct. English Nature: Peterborough, UK.

Frankland, J., Magan, N. & Gadd, G. M. (1996). *Fungi and Environmental Change*. Cambridge University Press: Cambridge, UK.

Homolka, L., Lisa, L., Stoytchev, I. & Nerud, F. (1999). Culture collections of basidiomycetes (CCBAS). A tool for *ex-situ* conservation of basidiomycetes. In *Abstracts, XIIIth Congress of European Mycologists*, Alcalá de Henares (Madrid) Spain, 21–25 September 1999, p. 53. Abstracts Volume: Alcalá de Henares.

Ing, B. (1993*a*). A provisional European red list of endangered macrofungi. *European Council for the Conservation of Fungi, Newsletter* **5**, 3–7.

Ing, B. (1993*b*). Towards a red list of endangered European macrofungi. In *Fungi in Europe: Investigations, Recording and Conservation* (ed. D. N. Pegler, L. Boddy, B. Ing & P. M. Kirk), pp. 231–237. Royal Botanic Gardens: Kew.

Ing, B. (1996). Red Data List and decline in fruiting of macromycetes in relation to pollution and loss of habitats. In *Fungi and Environmental Change* (ed. J. Frankland, N. Magan & G. M. Gadd), pp. 61–69. Cambridge University Press: Cambridge, UK.

Ivancevic, B. (1998). A preliminary red list of the macromycetes for Yugoslavia. In *Conservation of Fungi in Europe*, Proceedings of the 4th meeting of the European Council for the Conservation of Fungi, Vipiteno (Sterzing), Italy, 9–14 September, 1997 (ed. C. Perini), pp. 57–61. Privately published, University of Sienna.

Janos, D. P. (1996). Mycorrhizas, succession, and the rehabilitation of deforested lands in the humid tropics. In *Fungi and Environmental Change* (ed. J. Frankland, N. Magan & G. M. Gadd), pp. 129–162. Cambridge University Press: Cambridge, UK.

Keller, J. (1991). La protection des champignons. *Schweitzerische Zeitschrift für Pilzkunde* **69**, 229–234.

Kotiranta, H. (1998). A comparison between Finnish national threat categories and the 'new' IUCN threat categories; polypores as examples. In *Conservation of Fungi in Europe*, Proceedings of the 4th meeting of the European Council for the Conservation of Fungi, Vipiteno (Sterzing), Italy, 9–14 September, 1997 (ed. C. Perini), pp. 63–64. Privately published, University of Sienna.

Kotiranta, H. (1999). The threatened Aphyllophorales of Finland. In *Abstracts, XIIIth Congress of European Mycologists*, Alcalá de Henares (Madrid) Spain, 21–25 September 1999, p. 66. Abstracts Volume: Alcalá de Henares.

Leonard, P. (1997*a*). Towards a national policy on the commercial collecting of fungi. *Mycologist* **11**, 27–28.

Leonard, P. (1997*b*). A scientific approach to a policy on commercial collecting of wild fungi. *Mycologist* **11**, 89–91.

Lizon, P. (1999). Current status and perspectives of conservation of fungi in Slovakia. In *Abstracts, XIIIth Congress of European Mycologists*, Alcalá de Henares (Madrid) Spain, 21–25 September 1999, p. 77. Abstracts Volume, Alcalá de Henares.

Lonsdale, D. & Gibbs, J. N. (1996). Effects of climate change on fungal diseases of trees. In *Fungi and Environmental Change* (ed. J. Frankland, N. Magan & G. M. Gadd), pp. 1–19. Cambridge University Press: Cambridge, UK.

Magan, N., Smith, M. K. & Kirkwood, I. A. (1996). Effects of atmospheric pollutants on phyllosphere and endophytic fungi. In *Fungi and Environmental Change* (ed. J. Frankland, N. Magan & G. M. Gadd), pp. 90–101. Cambridge University Press: Cambridge, UK.

May, T. (1998). Protecting the forgotten kingdom: natural history collections and the conservation of fungi. [from a lecture given at the ICOM, International Council of Museums, 98, Melbourne, Australia, 13 October 1998]. <http://senckenberg.uni-frankfurt.de/icom/nh-melpr.htm>

O'Dell, T. E., Kiester T. & Molina R. (1999). Regional conservation strategies and distribution analyses of rare and old-growth forest dependent fungi in the Pacific Northwest, USA. In *Society for Conservation Biology 1999 Annual Meeting; 17–21 June at the University of Maryland, College Park. Integrating Policy and Science in Conservation Biology.* <http://www.inform.umd.edu/EdRes/Colleges/LFSC/FacultyStaff/dinouye/scbabstracts/Odell.htm>

Otto, P. & Ohenoja, E. (1998). Proposal and a preliminary conception for a project 'mapping of threatened fungi in Europe'. In *Conservation of Fungi in Europe*, Proceedings of the 4th meeting of the European Council for the Conservation of Fungi, Vipiteno (Sterzing), Italy, 9–14 September, 1997 (ed. C. Perini), pp. 145–148. Privately published, University of Sienna.

Palm, M. E. & Chapela, I. H. (1997). *Mycology in Sustainable Development: Expanding Concepts, Vanishing Borders*. Parkway Publishers: Boone, NC, USA.

Parmasto, E. & Parmasto, I. (1997). Lignicolous Aphyllophorales of old and primeval forests in Estonia. 1. The forests of northern Central Estonia with a preliminary list of indicator species. *Folia Cryptogamica Estonica* **31**, 38–45.

Pettitt, T. R. & Parry D. W. (1996). Effects of climate change on Fusarium foot rot of winter wheat in the United Kingdom. In *Fungi and Environmental Change* (ed. J. Frankland, N. Magan & G. M. Gadd), pp. 20–31. Cambridge University Press: Cambridge, UK.

Pilz, D. & Molina, R. (1996). *Managing Forest Ecosystems to Conserve Fungus Diversity and Sustain Wild Mushroom Harvests*. General Technical Report

PNW-GTR-371. USDA Forest Service, Pacific Northwest Research Station, Portland, Oregon, USA, 104 pp.

Pilz, D. & Molina, R. (1997). American matsutake mushroom harvesting in the United States: social aspects and opportunities for sustainable development. In *Mycology in Sustainable Development: Expanding Concepts, Vanishing Borders* (ed. M. E. Palm & I. H. Chapela), pp. 68–75. Parkway Publishers: Boone, NC, USA.

Pop, A. (1998). A short report on fungi conservation in Romania. In *Conservation of Fungi in Europe*, Proceedings of the 4th meeting of the European Council for the Conservation of Fungi, Vipiteno (Sterzing), Italy, 9–14 September, 1997 (ed. C. Perini), p. 141. Privately published, University of Sienna.

Rotheroe, M. (1995). Saving an historic lawn: conservation progress report. *Mycologist* **9**, 106–109.

Rotheroe, M. (1996*a*). Fungal conservation in Centenary Year. *Mycologist* **10**, 98–99.

Rotheroe, M. (1996*b*). Implications of global warming and rising of sea-levels for macrofungi in UK dune systems. In *Fungi and Environmental Change* (ed. J. Frankland, N. Magan & G. M. Gadd), pp. 51–60. Cambridge University Press: Cambridge, UK.

Rotheroe, M. (1998). Wild fungi and the controversy over collecting for the pot. *British Wildlife* **9:6**, 349–356.

Rotheroe, M., Newton, A., Evans, S. & Feehan, J. (1996). Waxcap-grassland survey. *The Mycologist* **10**, 23–25.

Rowe, R. F. (1997). The commercial harvesting of wild edible mushrooms in the Pacific Northwest Region of the United States. *Mycologist* **11**, 10–15.

Senn-Irlet, B. (1998). Regional patterns of fungi in Switzerland, consequences for the conservation policy. In *Conservation of Fungi in Europe*, Proceedings of the 4th meeting of the European Council for the Conservation of Fungi, Vipiteno (Sterzing), Italy, 9–14 September, 1997 (ed. C. Perini), p. 118–125. Privately published, University of Sienna.

Tortic, M. (1998). An attempt to a list of indicator fungi (Aphyllophorales) for old forests of beech and fir in former Yugoslavia. *Folia Cryptogamica Estonica* **33**, 139–146.

Venturella, G. (1999). The conservation of *Pleurotus nebrodensis* through its cultivation. In *Abstracts, XIIIth Congress of European Mycologists*, Alcalá de Henares (Madrid) Spain, 21–25 September 1999, p. 131. Abstracts Volume: Alcalá de Henares.

3

Conservation and management of forest fungi in the Pacific Northwestern United States: an integrated ecosystem approach

RANDY MOLINA, DAVID PILZ, JANE SMITH, SUSIE DUNHAM, TINA DREISBACH, THOMAS O'DELL & MICHAEL CASTELLANO

Introduction

The vast forests of the Pacific Northwest region of the United States, an area outlined by the states of Oregon, Washington, and Idaho, are well known for their rich diversity of macrofungi. The forests are dominated by trees in the Pinaceae with about 20 species in the genera *Abies*, *Larix*, *Picea*, *Pinus*, *Pseudotsuga*, and *Tsuga*. All form ectomycorrhizas with fungi in the Basidiomycota, Ascomycota, and a few Zygomycota. Other ectomycorrhizal genera include *Alnus*, *Arbutus*, *Arctostaphylos*, *Castinopsis*, *Corylus*, *Lithocarpus*, *Populus*, *Quercus*, and *Salix*, often occurring as understorey or early-successional trees. Ectomycorrhizal fungi number in the thousands; as many as 2000 species associate with widespread dominant trees such as Douglas-fir (*Pseudotsuga menziesii*) (Trappe, 1977). The Pacific Northwest region also contains various ecozones on diverse soil types that range from extremely wet coastal forests to xeric interior forests, found at elevations from sea level to timber line at 2000 to 3000 metres. The combination of diverse ectomycorrhizal host trees inhabiting steep environmental and physical gradients has yielded perhaps the richest forest mycota of any temperate forest zone. When the large number of ectomycorrhizal species is added to the diverse array of saprotrophic and pathogenic fungi, the overall diversity of macrofungi becomes truly staggering.

Issues relating to conservation and management of forest fungi in the

Pacific Northwest must be placed in the context of public land resources and Federal laws regulating forest management. The Pacific Northwest has the greatest amount of Federal forest in the United States, about 9.8 million hectares in Oregon and Washington alone. The lands are managed by the US Department of Agriculture (USDA), Forest Service, and the US Department of the Interior (USDI), Bureau of Land Management and National Parks Service. Each agency has different overall management goals. For example, National Parks do not allow commercial extraction of any forest products. National Park lands are primarily for recreational purposes and also serve as primary biological reserves. Other Federal forest lands vary between roadless wilderness areas with virtually no resource extraction to areas where forests are managed to meet multiple-resource objectives, including timber harvest, recreational access, and conservation of rare species. Several Federal laws and statutes, including the Endangered Species Act, the National Environmental Protection Act, and the National Forest Management Act, provide strict rules governing the planning and implementation of forest management and monitoring activities. These laws also provide for public input in planning and management. Over the last four decades, legal challenges to the development and implementation of forest plans have greatly shaped the current directions for managing public forests. These directions and decisions have direct relevance to the conservation and management of forest fungi.

During the late 1980s and early 1990s a series of lawsuits sharply curtailed the harvest of old-growth forests in the Pacific Northwest. At the centre of the controversy was the old-growth dependent northern spotted owl (*Strix occidentalis*), a rare species protected by the Endangered Species Act. Because the economy of many Pacific Northwest communities relied strongly on timber harvest from Federal forests, this action created political controversy. In 1992, President Clinton held a regional forest conference as a first step toward balancing the economic needs of the region with the protection of endangered species and old-growth forest habitat on Federal land. This eventually led to a scientific assessment of the problem (FEMAT, 1993) and development of the Northwest Forest Plan (USDA & USDI, 1994*a*). The plan strives to provide for a sustainable level of timber harvest and persistence of old-growth forest dependent species based on a region-wide allocation of land uses, including reserve and nonreserve areas. As part of the overall scientific assessment process, many rare species requiring old-growth forest habitat were analysed for protection under the land reserves and forest management guidelines. In the final analysis, 234 old-growth dependent, rare, fungal species were listed for additional protection under the Survey and Manage guidelines of the Record of Decision (USDA & USDI, 1994*b*). The Record of Decision is the final legally binding document that defines how the Northwest Forest Plan is to be implemented. The Survey and Manage guidelines detail the course of mitigations required by management to protect the species. Although not a part of the listing process for the Endangered Species Act, this action still represents the first Federal listing of fungal species for protection in the United States and highlights the need for a research programme addressing poorly understood conservation issues for forest fungi.

The Northwest Forest Plan initiated a significant paradigm shift in forest management from an emphasis on timber extraction to ecosystem management. Ecosystem management is a holistic approach to decision making that integrates biological, ecological, geophysical, silvicultural, and socio-economic information into land management decisions to preserve biological diversity and maintain ecosystem functioning while meeting human needs for the sustainable production of forest products and amenities (Bormann *et al.*, 1994; Jensen & Everett, 1994). Planning is typically conducted at broad scales such as watersheds or regional landscapes. Fungi play significant ecosystem roles in nutrient cycling, food webs, forest diseases, and mutualisms, and therefore significantly influence seedling survival, tree growth, and overall forest health. Ecosystem approaches to forest management provide an avenue for integrating our understanding of the biological and functional diversity of forest fungi into recent forest management objectives. It is our contention, and indeed the theme of this chapter, that an ecosystem-based approach provides the most effective means for conserving this vast array of essentially cryptic species.

As the numbers and diversity of fruit bodies of forest fungi were declining in Europe (Arnolds, 1991) and Japan (Hosford *et al.*, 1997), the last two decades in the Pacific Northwest witnessed a dramatic increase in the harvest of wild, edible mushrooms with export to Europe, Japan, and elsewhere. In 1992, nearly 2 million kg of edible mushrooms was harvested in the Pacific Northwest, bringing over US$40 million to the economy (Schlosser & Blatner, 1995). Forest managers and the public questioned whether such harvest levels were sustainable or could negatively impact persistence of harvested species (Molina *et al.*, 1993). The reproductive biology of these species, their habitat requirements, and effects of mushroom harvest and forest management practices on their continued productivity are poorly understood. Although commercial species are not rare, the conservation of them also lends itself to ecosystem management approaches and has prompted new lines of research.

The overall mission of the Forest Mycology Team of the USDA Forest Service, Pacific Northwest (PNW) Research Station, Corvallis, Oregon, is to provide basic information on the biological and functional diversity of forest fungi and to develop and apply this information in an ecosystem context to both conserve the fungal resource and sustain the health and productivity of forest ecosystems. The research is divided into three problem areas:

1. *Ecology, biology, and functional diversity of forest fungi* This area considers issues related to fungal taxonomy, systematics and phylogeny, understanding and documenting fungal communities in space and time, examining effects of natural and anthropogenic disturbances on the ecology of forest fungi, and exploring important ecosystem functions of forest fungi, particularly their roles in ecosystem resiliency and health.
2. *Conservation biology of forest fungi* This area develops fundamental approaches to determining population dynamics, develops habitat-based models to predict species presence and future potential habitat for rare fungi, develops methods to survey for rare species, and explores management options to conserve old-growth associated fungal species under the guidelines of the Northwest Forest Plan.
3. *Sustainable productivity of commercially harvested edible forest fungi* This area addresses challenges associated with measuring productivity of ephemeral fruit bodies, determines the effects of mushroom harvest and forest management practices on future production, and develops protocols to monitor sustainable productivity of the resource.

This approach addresses fungal resource issues at different scales (species, populations, communities, and landscapes) so that the results are relevant to ecosystem planning. The beneficial mutualistic mycorrhizal fungi are given primary attention because of their importance to forest trees. The remainder of this chapter outlines many issues in these areas, our research approaches and directions, and application of this research to the conservation and management of forest fungi. We begin by discussing basic research components, including ecosystem function, community ecology, and population biology. We conclude by focusing on the development of tools and concepts to apply this knowledge, including applications in inventory and monitoring. Although this research programme is unique to the Pacific Northwest region and the Federal policies on public lands, we believe it provides a relevant framework to approach fungal conservation issues at various scales and for different objectives in other parts of the world.

Ecosystem functions, fungal communities and populations

Ecosystem functions

Conserving and managing forest fungi requires an understanding of how fungi function in maintaining healthy ecosystems. Educating land managers about how management practices affect the reproduction, growth, and overall function of fungi through workshops and literature is vital to the success of integrating fungal concerns into forest management plans. In our discourse with managers, we emphasise and discuss four critical ecosystem functions as follows.

1. *Nutrient cycling, retention, and soil structure* Almost all fungi participate in nutrient cycling. This is best known for saprotrophs (decomposers), particularly those that degrade wood (cellulose and lignin). Two other roles are equally important, however, in this ecosystem function. First, fungal mycelia are large sinks for organic carbon and nutrients in the soil. In ectomycorrhizal-dominated forests, a significant portion of the carbon fixed by host trees can go to the mycorrhizal structures and fungal mycelium to support this symbiotic association (Smith & Read, 1997). Some fungal mycelia and ectomycorrhizal root tips turn over rapidly to return nutrients to the soil, whereas other fungal structures such as rhizomorphs retain nutrients in relatively long-lived structures or shunt it to other parts of the mycelium as some sections die. What is important is that fungi are major sinks for maintaining nutrients on site, thereby preventing nutrient leaching and loss from soils. Secondly, fungi exude polysaccharides similar to root exudates, and these 'organic glues' (Perry *et al.*, 1989) are important in creating and stabilising soil microaggregates and soil micropores. These soil structures contribute to soil aeration and water movement. Because most terrestrial organisms are strongly aerobic, this fungal contribution to soil microaggregation is critical for overall maintenance of soil health and productivity.

2. *Food webs* Although it is common knowledge that animals from molluscs to mammals (including humans) eat mushrooms, the magnitude of this food source and its ecological importance is underappreciated. The greatest biomass of fungi resides in hidden mycelia that explore soil and organic substrates, and it is here that the importance of fungi in food webs and soil fertility is revealed. Literally tens of thousands of microarthropod species live in forest soils, and about 80% of those are fungivores (Moldenke, 1999). Microarthropods provide the biological machinery to shred the vast quantity of organic matter in forest soils and prepare it for the final mineralisation processes carried out by microbes. Given the large amount of fungal hyphae in forest soils noted above, the magnitude of this soil food web process becomes readily apparent.

   Fruit bodies of fungi (mushrooms and truffles) are also vital food sources for many wildlife, especially forest mammals. Most small mammals depend on hypogeous fruit bodies (truffles) for a significant part of their diets (Maser, Trappe & Nussbaum, 1978). The Pacific Northwest is biologically rich with hypogeous fungi; several species produce fruit bodies year-round (Colgan *et al.*, 1999). This knowledge of food web connections directly relates to the preservation of birds of prey, such as the endangered northern spotted owl, that eat small mammals which themselves subsist almost entirely on hypogeous fruit bodies. Thus, from an ecosystem standpoint, in order for managers to protect these megafauna, they must understand the direct connection to a basic food web component, the fungi.
3. *Fungal plant pathogens* Typically we perceive disease-causing fungi as agents of destruction and gauge their impact by the volume of timber destroyed. Although this is an important consideration, ecosystem management goes beyond timber management by recognising diseases as important processes that create forest structure vital to other species and processes. In addition to providing organic matter for nutrient cycling, for example, standing dead snags or heart rot in live trees create needed habitat for primary cavity nesting birds and secondary cavity users. Forest gaps created in root rot pockets allow for development of understorey plants, contributing to both species richness and increasing browse for large mammals such as deer and elk. Schowalter *et al.* (1997) provide a framework for

integrating the ecological roles of pathogens in managed forests.

4. *Mycorrhizal mutualisms* The benefits of mycorrhizas to plants are well documented and include efficient nutrient uptake, especially phosphorus; enhanced resistance to drought stress; and direct or indirect protection against some pathogens. From an ecosystem standpoint, mycorrhizal benefits extend much further in space and time. In addition to participation in nutrient cycling and food web processes, mycorrhizal fungi provide linkages among plants that affect plant community development and resiliency. Our research has emphasised an understanding of the mycorrhizal specificity between potential plant hosts and fungal association, with direct relevance to the ability of plants to be connected by commonly shared mycorrhizal fungi (Molina, Massicotte & Trappe, 1992). Indeed, many fungi can form mycorrhizas with diverse plant species, including across plant families. Brownlee *et al.* (1983) and Finlay & Read (1986) showed that carbon (photosynthates) can move from a donor to a recipient plant when connected by a common fungus. Simard *et al.* (1997) later showed under field conditions that a net flow of carbon can move from a donor plant to a shaded recipient plant via mycorrhizal fungus linkages. Thus, understorey plants may receive partial nourishment from overstorey plants via a shared mycorrhizal fungus network.

   Interplant linkages are equally important for other ecological reasons. During early plant succession, pioneering plants maintain or build the mycorrhizal fungal community by supporting it with photosynthates. Late-seral plants can develop mycorrhizas with many of the same fungi supported by early-seral plants and thus benefit from the ongoing active mycorrhizal processes. We believe that the positive feedback interactions of these plant–fungal associations in space and time contribute to ecosystem resiliency after disturbance (Perry *et al.*, 1989; Molina *et al.*, 1992).

Fungal communities

Little is known about the impacts of forest management activities on fungal communities in the Pacific Northwest. To manage effectively for the diversity of forest fungi, it is necessary to identify the fungal species within forests, describe their community structure in space and time, and understand the impacts of both human-induced and natural disturbances on successional dynamics of forest fungi. This approach places fungi into the same context as that of understanding the dynamics of plant communities, a context relevant to forest managers. Data from fungal community studies provide the basis for addressing critical research topics including (1) conservation biology of rare fungal species; (2) fungi as important indicators of forest community dynamics; (3) sustainability of edible, commercially harvested fungi; and (4) development of habitat models for fungi. In this section, we describe some of the primary considerations in setting objectives for the study of fungal communities and then describe our current approaches to determining fungal communities in forests of the Pacific Northwest. Examples of a few studies are given.

Description of fungal communities

The first step in setting study objectives is defining what the phrase fungal community means. In its simplest definition, a community is a group of organisms sharing a particular place. Community structure is characterised by the numbers of species present, their relative abundance, and the interactions among species within the community. Characteristics of communities emerge from historical and regional processes. Thus, it is important to define clearly the time and space boundaries of the study. For example, the problem of determining changes in fungal species on a recently dead tree or log differs from problems associated with examining the succession of forest floor fungi over several hundred years of forest development.

If specific functional groups or processes are selected for study, then a guild concept can help to focus the community definition. A guild is a group of species that uses various resources or shares a common function, such as fungi that form ectomycorrhizas or that decompose wood. Studies of fungi, plant host, and environmental relations often target a particular guild. Because ectomycorrhizal fungi have been the focus of most fungal community studies in the Pacific Northwest, they receive primary attention in this section.

Several factors influence mycorrhizal community structure, including plant species and habitat conditions, such as tree composition, tree age, disturbance, and soil nutrients (Molina *et al.*, 1992; Vogt *et al.*, 1992; Gehring *et al.*, 1998). The objectives for a community study might examine one causative factor or a combination of factors for effects on community development. The selection of factors strongly influences the study design and sampling intensity. For example, if one is interested in the effects of coarse woody debris (CWD) on ectomycorrhizal community structure, then a simple design of sampling fungi at different distances from large wood in similar forest stands might suffice. However, if one is also interested in the importance of CWD at different stages of forest development, or in forests of different tree densities, then a complex design with different forest treatments is needed, which increases sampling units, time, and expense.

As for most biotic communities, disturbance phenomena often trigger the successional dynamics of fungal communities. For example, changes in plant species composition from forest succession or large-scale disturbances significantly affect ectomycorrhizal species composition and total fruit body production (Dighton & Mason, 1985; Termorshuizen, 1991; Vogt *et al.*, 1992; Amaranthus *et al.*, 1994; North, Trappe & Franklin, 1997; Waters *et al.*, 1997; Baar, Horton & Kretzer, 1999; Colgan *et al.*, 1999). Fire and timber harvest are the two main disturbance events in Pacific Northwest forests. Several regional studies have shown a reduction in fruit body production or levels of ectomycorrhiza formation after forest thinning or fire (Harvey, Larsen & Jurgensen, 1980; Parke, Linderman & Trappe, 1984; Pilz & Perry, 1984; Waters, McKelvey & Zabel, 1994; Colgan *et al.*, 1999). Severity and length of time since disturbance, however, are key factors in evaluating the effects of fire or tree death on the ectomycorrhizal community (Visser, 1995; Miller, McClean & Stanton, 1998; Horton, Cazares & Bruns, 1998; Jonsson, Dahlberg & Nilsson, 1999; Stendall, Horton & Bruns, 1999; Baar *et al.*, 1999) and should be considered in examining fire disturbance effects on fungal community dynamics.

Ectomycorrhizal community analysis

Once the scope and objectives of the study are clearly defined, the appropriate options for sampling the fungal community are considered. Approaches to ectomycorrhizal community analysis include collection of fruit bodies, morphological typing of mycorrhizas, and molecular identification of fungal symbionts. Each approach differentially estimates species dominance. For example, fruit body studies address food resource questions about mammal mycophagy based on fruit body biomass. Identification of mycorrhizal symbionts on root tips best addresses the fungal community transporting nutrients to the host or fungal symbionts available for microarthropod mycophagy. Selection of an approach depends on the objectives of the study. Each approach has strengths and weaknesses in regard to sampling intensity, skills needed, and accuracy in assessing species occurrence, especially at broad spatial scales. These approaches are not mutually exclusive, and most researchers are using a combined approach.

Fruit body surveys poorly reflect the composition of subterranean ectomycorrhizal communities, thereby making interpretations difficult at the community level (Danielson, 1984; Gardes & Bruns, 1996; Dahlberg, Jonsson & Nylund, 1997; Kårén *et al.*, 1997; Gehring *et al.*, 1998; Jonsson *et al.*, 1999). Inherent problems with fruit body surveys include the sporadic production of fruit bodies, infrequent surveys, and the omission of easy-to-overlook fruit bodies such as resupinate forms found on the underside of woody debris.

Identification of ectomycorrhizal root tips to broad morphological types (Zak, 1973) became a popular alternative to fruit body collection for defining ectomycorrhizal communities. This method, refined by Agerer (1987–1998), Ingleby, Mason & Fleming (1990), and Goodman *et al.* (1996), has been applied to glasshouse soil bioassays, as well as field studies. Morphological typing of mycorrhizas is time consuming and taxonomic resolution is low.

More recently, the application of molecular techniques, in particular PCR–RFLP (polymerase chain reaction–restriction fragment length polymorphism) analysis and DNA sequencing, to mycorrhizal community research has improved our ability to identify morphological types to family, genus, or species level (Gardes *et al.*, 1991; White *et al.*, 1990; Gardes & Bruns, 1993; Bruns *et al.*, 1998). Although not a panacea, PCR-based methods significantly improved the study of ectomycorrhizal communities and have enhanced our ability to investigate fundamental aspects of ectomycorrhizal communities in several forest types (Gardes & Bruns, 1996; Dahlberg *et al.*, 1997; Gehring *et al.*, 1998; Horton & Bruns, 1998; Taylor & Bruns, 1999). They allow comparison of the observed diversity of fruit bodies with that of root tips and allow us to study known species that do not grow in culture. Additionally, molecular techniques are useful for taxonomic placement of fungal symbionts to the family level in forests where species fruit infrequently or not at all. Similar to challenges with morphotyping, studying the mycorrhizal community with molecular techniques is time consuming and may require restricting sample size. Restricted sampling might inadequately reflect the wide variances typically seen in ectomycorrhizal root tip community studies.

Current fungal community studies in the Pacific Northwest

Given the size of the region, the diversity of forest types, and the long-term nature of fungal community studies, we have focused on key forest habitats or forest management issues that affect not only fungi but the overall ecosystem management as well. This approach provides the best opportunity to interpret mycological findings and integrate results into forest management plans. It is also cost effective because large-scale forest manipulations are often prohibitively expensive for stand-alone mycological studies.

Because conservation of old-growth forests is a major concern in the Pacific Northwest (only about 10% of the original old-growth forest remains), we have compared the ectomycorrhizal fruit body diversity (both mushroom and truffle) in old-growth (>400 years old) to younger stands (30–60 years old) of Douglas-fir. We have found a similar number of species in old-growth compared to younger stands. However, the species composition differs, and the number of unique species is greater in old-growth versus younger stands (J. Smith and others, unpublished data). These results provide evidence that some fungal species require old-growth forest habitat and that protecting the remaining old-growth forests is important for maintaining fungal species diversity.

Many forest management experiments in the Pacific Northwest focus on accelerating development of old-growth structure by using various silvicultural approaches such as thinning, green-tree retention, and prescribed burning to return forest tree composition to that found before European settlement and wide-scale fire suppression. We currently have several integrated studies under way in large-scale forest management experiments that examine the effects of these management treatments on fungal diversity and function. One, done in conjunction with the Ecosystem Study in the state of Washington (Carey, Thysell & Brodie, 1999), is designed to manipulate forest stands 60 to 80 years old to hasten their return to old-growth structure and provide habitat for the northern spotted owl (*Strix occidentalis*); the owl is currently absent from these experimental study sites. The experiment examines silvicultural effects on the diversity of important food web functions. The focus of this study is on the truffle community because this is the primary food base for small mammals, which in turn are the main prey of the northern spotted owl. Our results (Colgan, 1997; Colgan *et al.*, 1999) describe the community structure of dominant truffle species in the treated stands and in small-mammal diets. A key finding was that truffle species dominance shifts under different forest thinning regimes. Within these managed stands, several truffle species fruit year-round, thereby providing an essential food base during winter months. These species can thus be considered keystone species in the ecosystem food web. This type of fungal diversity study allows managers to understand the role of maintaining fungal communities and how specific silvicultural treatments affect their diversity, abundance, and function.

Many forests in the Pacific Northwest and throughout the Western United States are experiencing decline, primarily because of fire suppression and silvicultural management. Decades of fire suppression and selective tree harvest have yielded dense forests of shade-tolerant tree species prone to water stress and catastrophic insect disease outbreaks (Agee, 1993). Forest managers are experimenting with prescribed fire and thinning to improve overall forest health. These experiments also lend themselves to investigating effects on the fungal community. J. Smith and others (unpublished data) are examining the effects of burning and thinning on the fungal community in ponderosa pine (*Pinus ponderosa)* forests. By using primarily PCR–RFLP techniques to document changes in ecto-mycorrhizal fungal community structure they found that fire communities are dominated by few ectomycorrhizal species. Such species may be important to ecosystem resiliency after fire disturbance.

In addition to the studies mentioned above, Pilz & Molina (1996) describe many of the current efforts in the Pacific Northwest aimed at documenting the biological diversity, function, and community dynamics of forest fungi. Additional studies are also under way to examine management treatments such as variable thinning, presence of large woody debris, and soil compaction on fungal diversity and shifts in species dominance. By understanding the effects of various management practices on fungal communities, forest managers can integrate fungal diversity and function into forest management planning.

Population biology

An effort to conserve species, especially rare species, must include some understanding of the population structure and dynamics of the species in question. Our understanding of the population biology of fungi, however, lags far behind that of other taxa for many reasons described below. To deal with the issue of rare fungal species within the Northwest Forest Plan, we thought it important to begin a research programme that explores ways to examine the population structures representative of forest fungal species. Such exploratory research requires large sample sizes not easily obtained from populations of rare taxa so the initial research projects described below are focused on species that are common but represent different life history strategies found in ectomycorrhizal fungi. As techniques become refined, we will address similar questions with rare species.

Two long-term goals of our population genetic research on ectomycorrhizal fungi are to gain some understanding of how past and current land management practices in the Pacific Northwest have altered the dynamics of fungal populations and to use this information to predict how future land management scenarios may affect persistence of fungal populations. Achieving these goals requires an understanding of the spatial and temporal scales at which evolutionarily significant processes occur within and among fungal populations, which processes are most important at spatial scales relevant to management, and how spatial heterogeneity in landscape structure affects these processes (Wiens, 1989, 1997).

Problems inherent to the study of fungal populations at forest landscape scales

Meaningful analysis and interpretation of genetic spatial patterns requires that the extent of any selected study area be large enough to encompass multiple units (discrete populations) affected by gene flow and that individual samples (grain) be dispersed such that variance of process effects within each unit can be quantified (Wiens, 1989). Several ecological characteristics of fungi impede our ability to select scales appropriate for quantifying the relative importance of genetic drift, natural selection, and gene flow within and among populations. Selecting an appropriate extent and grain for population genetic studies requires an understanding of how individuals are distributed across the landscape and how much area discrete populations can cover. This presents an obstacle with fungi because the mass of thread-like mycelium (thallus) that gives rise to fruit bodies is frequently hidden from view. Attempts to characterise the size and distribution of discrete thalli have identified individuals encompassing hectares (Smith, Bruhn & Anderson, 1992), square metres (Baar, Ozinga & Kuyper, 1994), and square centimetres (Gryta *et al.*, 1997). Dahlberg & Stenlid (1990) have also shown that ecological factors such as forest age can influence the size of individuals and alter the mechanism (sexual or vegetative) by which new thalli are established. The range and variance of individual size and distribution unique to each fungal species will dictate the scale at which population boundaries should be encountered, and set the minimum size limit for study area extent.

When sampling fungi in the field, it is impossible to discern what proportion of the fruit bodies collected represent genetically identical individuals, and genetically distinct mycelium masses may produce many fruit bodies simultaneously. Discrete thalli can be established either by sexual reproduction (genets), selfing, or clonal vegetative propagation (ramets) (Todd & Rayner, 1980); evolutionary interpretations hinge on correct estimates of the contribution of each reproductive type to the genetic diversity observed within populations (Avise, 1994). Estimating proportional contributions of individual genets and ramets to the total genetic diversity in a population may not be intuitive. Depending on the mating system of the species, a maximum of 25% or 50% of the monokaryotic spores produced by an individual fruit body can potentially grow and fuse to form genetically novel dikaryotic genets so closely related to the parent mycelium that the two may fuse (Smith *et al.*, 1992). This type of fine-scale genetic structuring coupled with variance in genet size can bias allele frequency and gene flow estimates such that potential dispersal distances are underestimated. Results that are misleading owing to lack of knowledge of fungal reproductive behaviour can lead to poor development of management options.

The conceptual foundation of our research approach

Determining the appropriate scale at which to ask management-oriented questions about fungal population dynamics requires the design of a nonarbitrary, objective method to determine the scales at which important evolutionary processes produce particular genetic spatial patterns. Observations from a mismatch of spatial scales (scale of measurements differs from scale at which the process of interest is acting) can result in erroneous identification of pattern–processes relations because the types or relative influences of processes may change as new spatial domains are entered (Wiens, 1989). For example, at fine scales, biotic factors such as inter- and intraspecific competition, forest stand age, species composition of understorey shrubs, and type and severity of disturbance history may be strongly correlated with patterns of genetic variability. At broad geographic scales, abiotic factors such as climate, rainfall patterns, exposure, and elevation may strongly influence patterns of genetic variability. Fine-scale processes may exert control over the total amount of genetic variability on which large-scale processes can act, and thus may be more relevant to management issues.

Delimitation of fungal population boundaries and identification of dispersal barriers require the development of sampling techniques that facilitate accurate gene flow estimates at forest stand, watershed, and regional spatial scales. These sampling techniques must also account for heterogeneity in landscape cover, disturbance type, and disturbance intensity at each scale. The ultimate goal of such a multiscale sampling regime is to ensure that the scale of measurements and the response of fungal populations to some variable of interest (for example disturbance) fall within the same spatial domain. Variability across fungal species in individual size and distribution, mating system, and dispersal capability makes it impossible to apply a single sampling protocol to all species. Long-term advancement in understanding evolutionary processes critical to maintaining genetic variability in fungal populations requires basic research on how genetic variability in species with different life histories is partitioned at different spatial scales. This fundamental research will lay a general foundation for predicting how land management strategies may affect rare species depending on their life histories. Multiscale genetic studies will also help define scales potentially appropriate for future research on demographics or habitat requirements.

Current research examples

Building a strong knowledge base requires that basic genetic research be done on model species that will yield large sample sizes and exhibit reliable fruit body production from year to year. One example is the distribution of genetic variation across different spatial scales in three model genera: *Cantharellus*, *Rhizopogon*, and *Suillus*. Research on such common species will provide initial studies with the statistical power sufficient to correctly identify evolutionarily important pattern–process relations (Dizon, Taylor & O'Corry-Crowe, 1995). Golden chanterelles (*Cantharellus formosus*) are an economically important, epigeous fungus that forms mycorrhizal associations with various host species over a broad geographic range (Molina *et al.*, 1993). *Rhizopogon vinicolor* is an ectomycorrhizal species in the Boletales that forms hypogeous, truffle-like fruit bodies but is closely related to epigeous boletes in the genus *Suillus*, including *S. lakei* (Bruns *et al.*, 1989). Both *R. vinicolor* and *S. lakei* are common and distributed throughout the range of Douglas-fir. All three fungal species fruit predictably and abundantly and provide ideal model species with which to initiate base-line genetic studies because they represent diverse life histories in respect to spore dispersal. The epigeous fruit bodies of *Cantharellus formosus* can remain on the landscape continuously producing basidiospores for over a month (Largent & Sime, 1995), whereas *Suillus lakei* mushrooms are more ephemeral. Assuming that wind is the primary mechanism of spore dispersal for these two species, the potential for long-distance wind dispersal over time is most likely lower in *S. lakei* than in *C. formosus*. In contrast, *R. vinicolor* is primarily dispersed by small-mammal mycophagy

(Maser, Maser & Trappe, 1985) and might be more spore dispersal limited than the two wind-dispersed species. Research on these three species is being undertaken currently by independent scientists in study areas that overlap so that results will be comparable across the life-history types.

Characterisation of genetic patterns at different spatial scales requires multiple, independent, hypervariable markers that measure variation within the nuclear genome of the organism of interest (Bossart & Prowell, 1998). Comparative studies of population-level markers have shown that the use of short, tandemly repeated DNA sequences (microsatellite repeats) is an effective method for detecting variability where other genetic markers fail (Hughes & Queller, 1993). These valuable molecular markers do not exist for any ectomycorrhizal basidiomycete fungus but are under development for these three focal species.

At landscape scales, patterns of genetic isolation and indirect estimates of gene flow among populations are calculated by using estimates of variance in allele frequencies within and among sampling units (Wright, 1951; Slatkin & Barton, 1989). Precision of these statistics requires that sampling units do not encompass multiple random breeding units (genetic neighbourhoods) (Magnussen, 1993; Ruckelshaus, 1998). Wright (1969) showed analytically that as more neighbourhoods are included within sampling units, the genetic variance among sampling units relative to the total variance declines, and the power to detect spatial genetic structuring is lost. Progressing to landscape-scale studies requires that we understand the scale at which random mating occurs in these species. Spatial autocorrelation analysis of genotype locations within forest stands allows estimation of the diameter of areas where random mating occurs (that is, genetic neighbourhoods) via description of isolation by distance patterns over the space that a species occupies (Sokal & Oden, 1991; Epperson, 1993). The information necessary for spatial autocorrelation studies includes mapped locations of fruit bodies collected over a broad range of distances, the distance between every pair of sample points, and the genotypes of each fruit body at multiple, independent loci. Test statistics are calculated for the null hypothesis that the genotypes are randomly located across the study area. Spatial autocorrelation studies are currently under way for *C. formosus*, *S. lakei*, and *R. vinicolor* in overlapping study areas.

Results of this research will aid in the design of a sampling method that will capture informative variability in genetic patterns and processes at spatial scales relevant to management. Future research efforts will apply the sampling techniques, currently in development, to the study of how heterogeneity in landscape features (for example forest stand age, patterns created by different disturbance regimes) affects rates of gene flow between populations of fungi. Using this line of research to develop a better understanding of how fungi disperse across landscapes will facilitate identification of geographic areas important to the conservation of genetic diversity.

Applications in inventory and monitoring

Identifying and prioritising rare species

Protecting rare species is often a prominent component in conservation programmes, regardless of the scale of the conservation effort. Resources for study of species biology are scarce, and society places a high value on maintaining rare species. Understanding the causes and consequences of species rarity is important because it is a basis for setting conservation priorities. Thus, it is important that mycologists identify fungal species that are most rare or at risk as one basis for allocating resources to studying particular species.

As noted in our introduction, the Survey and Manage guidelines of the Northwest Forest Plan provide protection of rare, old-growth forest dependent fungi. The species list was compiled by three expert regional mycologists, Joseph Ammirati, William Denison, and James Trappe, and was based on their extensive experience and personal records from over 30 years of collecting macrofungi throughout the region. The list includes gilled, cup, polypore, and truffle fungi (Table 3.1). Within the Survey and Manage programme, each listed species represents a set of hypotheses. Each species is thought to be (1) rare, (2) associated with old-growth forests or legacy components of old-growth forests such as coarse woody debris and (3) not adequately protected by the reserve land allocations and management actions to protect needed habitat. The Survey and Manage programme of work addresses these three hypotheses so that appropriate protection measures can be taken to assure species persistence.

The Survey and Manage conservation programme provides protection only for rare, old-growth forest associated species; we will address some of the unique considerations in conducting extensive surveys for those species in the next section. Two other conservation approaches also have been useful for identifying and prioritising rare fungal species in the Pacific Northwest, Natural Heritage programmes and Rarity, Endangerment, and Distribution lists (RED lists – hence, Red Data lists). These programmes are operated at state levels and cover all species, geographic

Table 3.1. *Fungal genera and number of species listed under the Survey and Manage guidelines of the Northwest Forest Plan*

Genus	Number of species	Genus	Number of species
Acanthophysium	1	*Hebeloma*	1
Albatrellus	4	*Helvella*	4
Alpova	2	*Hydnotrya*	2
Arcangeliella	3	*Hydnum*	2
Asterophora	2	*Hydropus*	1
Baeospora	1	*Hygrophorus*	3
Balsamia	1	*Hypomyces*	1
Boletus	2	*Leucogaster*	2
Bondarzewia	1	*Macowanites*	3
Bridgeoporus	1	*Marasmius*	1
Bryoglossum	1	*Martellia*	4
Cantharellus	3	*Mycena*	5
Catathelasma	1	*Mythicomyces*	1
Chalciporus	1	*Neolentinus*	2
Chamonixia	1	*Neournula*	1
Choiromyces	2	*Nivatogastrium*	1
Chromosera	1	*Octavianina*	3
Chroogomphus	1	*Omphalina*	1
Chrysomphalina	1	*Otidea*	3
Clavariadelphus	6	*Phaeocollybia*	13
Clavicorona	1	*Phellodon*	1
Clavulina	2	*Pholiota*	1
Clitocybe	2	*Pithya*	1
Collybia	2	*Plectania*	2
Cordyceps	2	*Podostroma*	1
Cortinarius	13	*Polyozellus*	1
Craterellus	1	*Pseudaleuria*	1
Cudonia	1	*Ramaria*	28
Cyphellostereum	1	*Rhizopogon*	12
Dermocybe	1	*Rhodocybe*	1
Destuntzia	2	*Rickenella*	1
Dichostereum	1	*Russula*	1
Elaphomyces	2	*Sarcodon*	2
Endogone	2	*Sarcosoma*	2
Entoloma	1	*Sarcosphaera*	1
Fayodia	1	*Sedecula*	1
Fevansia	1	*Sowerbyella*	1
Galerina	5	*Sparassis*	1
Gastroboletus	5	*Spathularia*	1
Gastrosuillus	2	*Stagnicola*	1
Gautieria	2	*Thaxterogaster*	2
Gelatinodiscus	1	*Tremiscus*	1
Glomus	1	*Tricholoma*	1
Gomphus	4	*Tricholomopsis*	1
Gymnomyces	2	*Tuber*	2
Gymnopilus	1	*Tylopilus*	1
Gyromitra	5		

locations, and habitat types within the state (see, for example, Washington Natural Heritage Program, 1997). Both approaches use similar criteria to evaluate species and ranking systems that assign risk status. Species evaluation criteria include pattern of occurrence in space and time, vulnerability of habitat, direct and indirect threats to species decline, degree of protection under current management guidelines, and taxonomic uniqueness. Rankings are based on number of occurrences (rarity), number of individuals, population and habitat trends, and type and degree of threats. Status categories under the natural heritage programme include: endangered, threatened, sensitive, and possibly extinct or extirpated in the state. Two other categories, review and watch, are assigned to species of uncertain status for which additional information is needed before it is assigned to one of the four previous categories.

Red Data lists serve several purposes for fungi (see Arnolds, 1989). They alert mycologists and conservation biologists to issues surrounding rare fungal species and provide information to managers and policy makers responsible for the management and protection of the species. Some of the listed species are useful as candidates for long-term monitoring of status and trends of fungal populations. Scientists can also compare lists from different geographical areas to estimate national and international status of individual species.

Nearly 25 years ago the first reports concerning a decrease in fungal species diversity and the abundance of fungal fruit bodies across a wide geographic area originated from Europe (Schlumpf, 1976; Bas, 1978). This awareness led to creation of Red Data lists in 11 different European countries in the next 15 years (Arnolds & De Vries, 1993). Attention to this phenomenon outside northern Europe was lacking until recently.

In conjunction with the species information generated by the Northwest Forest Plan process, another regional assessment on the ecosystems of eastern Oregon and Washington, and the western parts of Idaho (called the interior Columbia River basin assessment; see USDA & USDI, 1996) led to the inception of the first Red Data lists of macrofungi for any large geographic area outside of northern Europe (Castellano 1997*a*, 1997*b*). The Red Data list for Oregon contains 118 species of macrofungi. It is smaller than the Survey and Manage list for Northwest Forest Plan lands because it evaluates species across the entire state and across all habitat, not just late-successional forest habitat. Some species that are rare on protected Federal land are sometimes more common in younger stands or on other land ownership. The interior Columbia River basin assessment yielded a Red Data list of macrofungi for Idaho (Castellano, 1998) containing nearly 200 fungal species of special concern. Simultaneously, J. Ammirati developed a working list of 57 rare macrofungi for Washington (Washington Natural Heritage Program, 1997). These three listing efforts are considered preliminary. Additional systematic and strategic sampling of habitats is critical to uncover previously undiscovered occupied sites for many of these species. Significant efforts are under way by the USDA Forest Service and USDI Bureau of Land Management to collect these fungi in these regions.

As more fungal surveys occur, these lists will likely change. It is important that mycologists update Red Data lists and other lists, such as those for Natural Heritage programmes, and disseminate the information so that the mycological community and the public can stay abreast of ongoing efforts in cataloguing rare species, prioritising them into risk categories, and monitoring our success in providing for their conservation. As lists are developed for different parts of the world, we can better understand distribution patterns of rarity and key habitats needed for their protection. We also can begin to understand some of the natural and anthropogenic factors that lead to fungal species rarity.

Methodological considerations in fungal inventories

The Survey and Manage guidelines of the Northwest Forest Plan centre around the adaptive management principle of first providing for species at known sites by appropriate management of habitat (manage) and simultaneously searching for new locations (survey) to learn more about the rarity, range, distribution, and habitat requirements. Gathering range and habitat information for over 200 rare fungal species in a region the size of the Northwest Forest Plan is unprecedented in scope for any fungal inventory programme. Originally, the surveys were to be conducted for at least 10 years to collect enough new information to develop science-based management plans for maintaining species persistence. Given the above survey objectives, several considerations needed to be addressed for effectively implementing regional-scale surveys. This section highlights several fundamental areas that we considered in conducting broad-scale fungal surveys and inventories.

Taxonomic expertise

Taxonomic knowledge is the cornerstone of biology and is essential in efforts to conserve biodiversity. Fungal taxonomy is problematic because our understanding of species is limited; Hawksworth (1991) estimated that less than 5% of the mycota has been described in the scientific literature. Even in well-known and economically important fungal groups, new species are continually described and some species concepts are in flux (e.g. Redhead, Norvell & Danell, 1997; Hughes *et al.*, 1999; Sime & Petersen, 1999).

Taxonomists are scarce because of a shift in academic programmes toward molecular systematics and ecology. The depleted ranks of classical fungal taxonomists can be augmented, however, by a cadre of experienced parataxonomists, people with less formal schooling in mycology, who are trained and gain significant experience in fungal identification. Over the past three years, we have trained many Federal agency biologists and botanists in the proper identification and handling of fungal specimens. Such training is essential to conduct the regional surveys. Training also increases the awareness of field staff and managers about the natural history of fungi.

Even when accurate identifications are performed, observations of species occurrences are anecdotes unless a voucher specimen is deposited in a publicly accessible herbarium (Ammirati, 1979). Properly documented voucher collections allow experts to confirm the identification of potentially rare species. We have entered into formal partnerships with regional university herbaria to access the many fungal specimens collected from regional surveys.

Survey approaches

Biological surveys need clearly defined goals. This is also the case when surveying for both rare and potentially rare species. Clearly stated goals and prioritisation of those goals allow one to define which data to collect, where and when to survey, and which species to collect. One can prioritise by species, ranking from most to least important, or by geography, ranking different habitats or regions, for example, by where the most species are expected to occur, or by where there are the least historic data. In practice, with planning and organisation, multiple goals may be pursued simultaneously. For example, one can visit sites of the rarest species, and while in the vicinity, survey high-priority habitats.

Prioritising species for survey Ranking species according to rarity is probably the most frequent approach to conservation prioritisation. We use most of the ranking criteria developed by the natural heritage programmes mentioned previously. These criteria evaluate the rarity and distribution of species, with the highest priority assigned to potentially extinct species, followed by critically imperilled species (typically known from five or fewer locations). The criteria are applied at both the global and local (state) scale, so a species can be extinct in the state, but globally secure. The rank also is evaluated as to the certainty with which it can be applied, and the certainty of taxonomic status.

Once species are prioritised, the task of locating begins. Either historic sites or potential habitat can be targeted; both approaches can be challenging. The rarest species are likely to be known only from historical records that may lack the precision necessary to find the site. Similarly, habitat for rare species is usually poorly defined. Two important goals for visiting historic sites include documenting whether the species still occurs there, and gathering habitat data. An additional bonus comes, not uncommonly, when other rare species are found at such historic sites.

Prioritising regions for survey One approach to regional prioritisation is determining appropriate habitat. In many cases, however, species habitat preferences are not known. Herbarium labels are notoriously vague about habitat, with such descriptors as 'in woods', 'under pine', or 'in conifer forests'. Nevertheless, for some species there are relatively narrow host or habitat requirements that can guide one to areas of interest. For example, *Gelatinodiscus flavidus* is a rare ascomycete that fruits only on the foliage of *Chamaecyparis nootkatensis*, a conifer with a fairly restricted distribution. Another example is the number of species on the Survey and Manage list that fruit in montane forests, usually including *Abies*, near melting snowbanks. Where specific habitats can be identified, prioritising surveys in those habitats can often extend the documented range of a species.

A second approach that we have used with good success is visiting data-poor areas. The existing regional database of known site locations is analysed to identify a data-poor area, that is an area with few known sites of targeted rare species. After gathering herbarium data for about 135 putatively rare species of fungi, it became apparent that much of the data resulted from a few mycologists visiting a restricted set of collecting areas (Castellano & O'Dell, 1997; Castellano *et al.*, 1999). The data were obviously biased, raising the question of how to use it to best advantage. We have used GAP analysis following the procedures of Kiester *et al.* (1996) to analyse regional distribution of rare fungi site locations and select priority areas for survey. This approach has the advantage that when observations of a species of interest are made in data-poor areas, they are likely to extend the known range or habitat amplitude for the species. Many new observations of Survey and Manage fungi have resulted from applying this strategy (Castellano *et al.*, 1999).

Survey techniques

After selecting where and for which species to survey, one must design an effective survey method. Again, survey methods must have clear objectives. For example, finding occurrences of rare species requires a different approach than testing a hypothesis about their habitat preferences. We have used two primary methods, controlled intuitive and systematic or random sampling methods.

Controlled intuitive survey Controlled intuitive survey is the simpler of the two methods. Once the site has been determined, surveyors travel to the site and survey the area, focusing on habitats or microsites where they think the target species is likely to occur. This approach relies on knowledge, experience, observation, and intuition, and may be the most reliable method for locating rare species.

Systematic or random sampling methods If information on relative abundance of species, or quantitative comparisons of different sites or habitats are needed, then methods that employ systematic random site and plot selection must be used that standardise survey intensity and meet statistical sampling assumptions. Plots that define a survey area are one way of standardising survey effort. Because fungi tend to occur in clustered patterns on the landscape, long narrow plots are often used; they are more efficient at sampling such distributions than circular plots (T. O'Dell, unpublished data). For truly rare species, however, random and systematic approaches may prove less useful than controlled intuitive surveys because of the low probability of encountering the species of interest. Adaptive sampling methods have been recommended for organisms that are patchily distributed and rare, as is the case for some fungi (Thompson, 1992). In this approach, sampling intensity is increased when the target species is encountered. For example, if a rare fungus is found, more plots or time could be added to the sample in the immediate or nearby vicinity. In this way, more useful information on its population size and habitat association can be gathered.

Limitations and inferences

Uncertainty surrounds the field sampling of fungi. No matter how much sampling is done, species will be undetected owing to seasonal and annual variation in fruiting, limitations to the amount of area scrutinised, and inconspicuousness of fruit bodies. Most macrofungi produce fleshy, ephemeral fruit bodies. Unless sampling occurs during the exact time it is fruiting, a species might go undetected. In the Pacific Northwest, fleshy fungi can be collected every month of the year in one habitat or another, but we tend to sample during limited times to make the most of available survey resources. Some species, therefore, are overlooked simply because they fruit at times when most other species do not, and when we are unlikely to be sampling. During peak fruiting seasons, sampling every two weeks may fail to detect a significant fraction of species compared with weekly sampling (Richardson, 1970). Unless an intensive study is being conducted, however, repeated visits to a site may be impractical. Annual variation in species occurrence can be tremendous, with as little as 5% of species observed at sites in one year recurring the next year (O'Dell, Ammirati & Schreiner, 1999). Many repeated visits to a site are required to determine the presence or absence of a species at that site.

Although we have noted several difficulties in conducting broad-scale, systematic surveys for rare fungi, it is important that mycologists are not discouraged from pursuing the need for fungal inventories. We stress again the importance of defining clear objectives for the surveys, designing the appropriate methods to collect the needed information, and scheduling enough time to conduct the surveys to satisfy the original objectives.

Modelling habitat as a tool in conserving fungal resources

Unlike conservation of rare plants and animals, which may entail reintroducing species into an area of suitable habitat as part of species recovery and conservation, the primary means of conserving rare fungal species requires protecting habitat and providing future habitat through land management planning. Currently, we understand poorly the specific habitat needs of most fungal species so it is difficult to develop land management options that provide for persistence of habitat and therefore the species. Even as we begin to collect habitat data from our regional surveys, it remains difficult to discern the key environmental and biotic habitat variables to which the fungi in question are responding. Because of this lack of knowledge, we have begun a fungal habitat modelling research project that will allow us to build suitable habitat indices, construct hypotheses around key variables, and then test the model for accuracy in predicting suitable fungal habitat. Because habitat modelling has rarely been attempted for fungi, in this section we first define habitat modelling and its uses, and then describe unique considerations for modelling fungal habitat and our research approaches.

Habitat modelling: definition and uses

A habitat model is an abstract representation of the physical and temporal space in which an organism lives, including the way it interacts with the surrounding biotic and abiotic environment (Patton, 1997). Habitat modelling can potentially be a powerful tool in the study of fungi and in fungal conservation biology for several reasons. Habitat modelling provides insight into the behaviour of a system and an understanding of the complex interaction between an organism and the environment. As a method for integrating data from diverse research areas, modelling also provides a formal organising framework for generating hypotheses, conducting data analysis, and assisting in determining research directions by helping to define problems and refine questions. Habitat modelling in particular can be used to develop ideas about the distribution and occurrence of forest fungi, thus contributing to the development of conservation strategies.

Land managers need to make decisions, regardless of whether they have sufficient data and understanding of ecosystems. By understanding habitat, we will be able to map geographic areas where particular fungal species are likely to occur. Once occurrence and species habitat relations have been established, simulation of habitat availability under different management scenarios and under various disturbances will allow projections of potentially occupied habitat over time. Although the Northwest Forest Plan mandates surveys for more than 200 species of fungi, limited resources are available for conducting those surveys. Models will help prioritise data-collection efforts by predicting areas where species exist or where more intensive sampling may be desirable (Kiester *et al.*, 1996). The GAP analysis mentioned as a rare species search prioritisation method illustrates a type of modelling that is helping to design regional survey strategies for forest fungi. Using models as ecosystem management tools helps managers to calculate risks of decisions and actions, and provides a method of accountability and support to land management decisions. This will be particularly helpful in developing sustainable harvest strategies for commercially valuable fungi and in predicting the impact of human intervention on fungal survival and productivity.

Habitat modelling: challenges and concerns

What are the habitat needs of a given fungal species? We assume that fungal species are tightly linked biologically to their habitats; that these habitats can be detected, measured, and quantified; and that we are able to detect the fungus itself. Some are habitat specialists, restricted to one or a few host species or a unique vegetation type; others are habitat generalists, associated with various host plants or vegetation types. Our work investigates habitat requirements and modelling for fungal species that exhibit contrasting adaptation strategies: for example, aerial spore dispersal for epigeous species versus small-mammal dispersal for hypogeous species.

Guidelines for habitat model development (US Fish & Wildlife Service, 1981) have been used to generate models for mammals, fish, and birds. Habitat evaluation procedures provide quantification of habitat based on suitability and total area of available habitat. Variables are selected according to three criteria: (1) the variable is important to survival and reproduction of the species, (2) there exists a basic understanding of the habitat–species relation for the variable, and (3) quantifiable data are available and the variable is practical to measure. Few studies of fungi, however, have directly assessed any of these criteria.

We know little about fungal ecology in general, and until recently, we did not have the tools that allowed us to define the individual, let alone populations (see section on 'Population biology'). Many fungal species disperse passively by wind; however, we do not know the probabilities associated with spores landing in suitable habitat and developing into a fungal individual. For many fungal species, nutritional mode is uncertain, thereby making it difficult to use 'food source' as a habitat variable. Taxonomic problems and database biases encumber the study of mycology, as mentioned in the previous section on survey difficulties. The issue of scale also arises when examining mycological data and attempting to define habitat variables. Mycologists often focus on the minutiae and are unaware of larger scale phenomenon that may affect fungal distribution.

Current modelling approaches

We are compiling information and formulating a concept of habitat for several species. For forest fungi, we define habitat as all factors (biotic and abiotic) affecting the probability that a fungus can inhabit, survive, and reproduce in a given place and at a given time. A review of the published literature reveals potential habitat factors that fall into four broad categories: vegetation, climate, topography, and soils.

Vegetation may be the primary factor contributing to ectomycorrhizal fungal habitat, because of host specificity. Molina & Trappe (1982) recognise three groups of ectomycorrhizal fungi based on the relations to host species, ranging from highly host specific to host generalists. Decomposer fungi also show various levels of host-substrate specificity (Swift, 1982). We can use, therefore, the presence or absence of particular host species as an initial indicator for potential habitat. In addition to host specificity, forest stand age and structure, and disturbances such as timber harvest and fire contribute significantly to fungal habitat (see section on 'Fungal communities').

Soil organic matter, including humus and coarse woody debris (CWD), is a substrate for ectomycorrhizal fungi (Harvey, Larsen & Jurgensen, 1976) and decomposer fungi. The CWD may be particularly important as a moisture-retaining substrate, thereby allowing root tips to support active ectomycorrhizas in times of seasonal dryness, on dry sites, or after fire (Harvey, Jurgensen & Larsen, 1978; Amaranthus, Parrish & Perry, 1989, Harmon & Sexton, 1995). Although no data exist for determining quantities of CWD necessary to support viability of forest fungi in the Pacific Northwest, availability of CWD seems to be a factor in establishing some fungi as well as seedlings (Kropp, 1982; Luoma, Eberhart & Amaranthus, 1996). The presence of CWD, therefore, may be associated with habitat for certain fungal species.

Topographic and soil factors important to fungal habitat include elevation, slope, aspect, soil properties, and local microtopography. To date, no studies have presented significant correlations with any of these factors, although mycologists often use such factors intuitively to find forest fungi.

Climate is a complex factor in fungal distribution and therefore habitat. Seasonal and ecological distribution of fungi as well as fungal productivity are partly determined by temperature and moisture (Norvell, 1995; O'Dell, Ammirati & Schreiner, 1999). Wilkins & Harris (1946) contend that moisture may be the most important single environmental factor controlling fungal reproduction. We know little, however, about the effects of climate on long-term survival of fungi or on timing of mushroom formation. Many scientific researchers, commercial mushroom harvesters and recreational forest users have much experiential knowledge as to when and where particular mushroom species are found. Knowledge- and rules-based modelling methods and expert systems capture such information; both quantitative and qualitative information can be synthesised (Starfield & Bleloch, 1991). In addition, the availability of digital maps and geographic information systems provide opportunities for linking our modelling rules with spatial databases and simulation models for other biological systems. Dreisbach, Smith & Molina (in press) provide more details on our current modelling approaches.

Sustaining the commercial harvest of edible forest mushrooms

As noted in the introduction, conservation of fungi entails more than protecting rare species. Conservation includes sustaining populations of all fungal species, both rare and common. As described by Arnolds (1991), common fungi can become rare because of habitat loss or large-scale anthropogenic disturbances such as air pollution. The commercial harvest of common, edible mushroom species illustrates a human disturbance that might diminish fungal resources. This section describes our research programme aimed at understanding this resource so that we can continue to enjoy economic gain and culinary delights while sustaining the survival and reproduction of these valuable fungi. We outline edible mushroom harvesting issues and concerns, discuss results from various recent research projects, describe forest management implications of the findings, and consider future directions in edible mushroom research and monitoring.

Scope of issues

Edible mushrooms have been widely collected from the forests of the Pacific Northwest since the 1860s when European settlers began hunting for mushrooms similar to species they had collected in their homelands. Some Native American tribes harvested mushrooms for various uses, but we lack evidence of widespread consumption (Kuhnlein & Turner, 1991). Regionally, amateur mushroom clubs and organised mushroom events date to the 1950s (Brown *et al.*, 1985), and now there are about 34 mycological societies based in the region (Wood, 2000). Academic mycological activities date back to the 1920s for the region. During the 1980s and early 1990s, commercial mushroom harvesting expanded dramatically. The increased public demand for wild edible mushroom-harvesting opportunities has affected all forest landowners in the Pacific Northwest, especially Federal forests.

Federal lands encompass a large portion of suitable mushroom habitat in the Pacific Northwest, and are managed for multiple use, including commercial forest products. Since the 1950s timber has been the major forest product, but environmental concerns and adoption of the ecosystem management philosophy defined in the introduction have led to dramatic

Table 3.2. *Important nationally and internationally marketed wild mushrooms harvested from the forests of the Pacific Northwest*

Local common names	Latin name	Use
American or white matsutake, pine or tanoak mushroom	*Tricholoma magnivelare* (Peck) Redhead	Food
Morels	Various *Morchella* species	Food
Pacific golden chanterelle	*Cantharellus formosus* Corner (previously called *C. cibarius*, Redhead *et al.* 1997)	Food
White chanterelle	*Cantharellus subalbidus* Smith & Morse	Food
Hedgehogs	*Hydnum repandum* L. ex Fr. *Hydnum umbilicatum* Peck	Food
King bolete	*Boletus edulis* Bull.:Fr.	Food
Oregon white truffle	*Tuber gibbosum* Harkn.	Flavouring
Oregon black truffle	*Leucangium carthusianum* (Tul. & C. Tul.)	Flavouring

reductions in timber harvesting from Federal lands during the 1990s. In response to reduced Federal timber harvests, many forest workers broadened the range of commercial products that they harvested from forests. Other social and ethnic groups joined them for various reasons (Arora, 1999; Love, Jones & Leigel, 1998; Sullivan, 1998). The mushroom industry has benefited from and been encouraged by the development of international markets and increased use of wild mushrooms by gourmet chefs (Schneider, 1999). The harvest of nontimber forest products in general has become a widely recognised industry in the Pacific Northwest and globally (Ciesla, 1998; Lund, Pajari & Korhonen, 1998). In 1992, nearly 11 000 people, employed full- or part-time, contributed over US$41.1 million to the economies of Oregon, Washington, and Idaho by harvesting and selling nearly 1.82 million kg (4 million pounds) of wild edible mushrooms (Schlosser & Blatner, 1995). Likely, the harvest has grown since then. Schlosser & Blatner (1995) state that over 25 species of mushrooms and truffles are commercially harvested in the Pacific Northwest. Table 3.2 lists the most economically important species. With the exception of some morels, all of the fungi in Table 3.2 are ectomycorrhizal.

Biological and management concerns

The magnitude of the commercial mushroom harvest in the Pacific Northwest has raised controversy about conservation of the mushroom resource, particularly regarding harvest effects on species viability and ecosystem function. As part of a comprehensive research programme for managing commercial mushroom harvests, we have conducted several public workshops to gather information on the primary concerns of resource managers, the mushroom industry, and the general public. Those concerns fall into four categories; we present them as a series of questions so readers can envision the scale and complexity of required mycological research.

Production and distribution

How many fruit bodies are being produced? How are they distributed across the landscape or within a given area? How does production differ during a season and from year to year? What is the actual or potential commercial productivity of a given area? What proportion of forest habitat is available and accessible for harvesting? What factors determine productivity and how might they be managed?

Mushroom harvesting

How can the sustainability of mushroom harvesting be assured? What proportion of the crop can be harvested without unacceptable impacts on the fungus itself or other resources? What techniques will mitigate those impacts; does mushroom harvesting increase or decrease subsequent production? Is spore dispersal reduced by removal of immature mushrooms, and, if so, does it impair reproductive success and degrade genetic variability? Is fungal mycelium and subsequent mushroom production affected by search and harvest techniques such as raking, moving woody debris, or digging? Do numerous harvesters trampling the forest floor harm mushroom production? How important are commercially valuable species as food for wildlife, and is human competition for the resource significant?

Forest land management

How do various timber harvesting methods (clear cutting, thinning, host species selection) affect subsequent mushroom production? What is the impact of soil compaction or disturbance from logging activities? What are the relations between fire and subsequent mushroom production, especially for morels? How does the intensity and timing of fire influence edible mushroom production? What impacts do grazing, fertilisation, or pesticide application have on mushroom production? Can mushroom production be improved through habitat manipulation (for example: planting tree seedlings inoculated with specific fungi, thinning understorey brush for sunlight and rainfall penetration, prescribed burning, or irrigation)? Can production be increased across the landscape by managing forests to attain tree age class, structure, and composition optimal for mushroom production?

Biology and ecology

What are the important reproductive events in the life cycle of a particular fungus species? How are new colonies or populations established and maintained? What causes them to diminish or perish? How important is spore dispersal to reproductive success, population maintenance, genetic diversity, or adaptability to unique microhabitats? How much genetic diversity exists within and among populations? Are there endemic, narrowly adapted, or unusual populations of otherwise common species? What are the growth rates of fungal colonies in soil and degree of mycorrhizal development by specific fungi on root systems? To what degree do other mycorrhizal or saprotrophic fungi compete with marketable fungi for colonisation sites on host roots or space in the forest soil?

Various measures have been adopted on Federal forest lands to ameliorate impacts until more is known about the long-term consequences of intensive commercial mushroom harvesting. Maintaining appropriate forest habitat is the most important and practical means of sustaining healthy populations of fungi. Essential habitat characters include appropriate host tree species for ectomycorrhizal fungi, coarse woody debris for wood inhabiting fungi and noncompacted soils for both saprotrophic and ectomycorrhizal fungi that fruit on the forest floor. Interim conservation measures that can be used until needed research and monitoring is conducted include no-harvest areas, fallow harvest rotations, limited harvest seasons or times, limited harvest permit numbers, limited harvest quantities, and incentives for leaving nonvaluable older specimens to disperse their spores. The approaches used in any given situation depend on land tenure, land management goals, mushroom values, availability of law enforcement, and especially, support for regulations among harvesters themselves.

Answering the questions above requires scientific investigations at various ecological scales (soil microniches to regional landscapes) and by using various investigative methods. Some of our approaches are noted below.

Research findings and management implications

In the Pacific Northwest, edible mushroom research is currently concentrated on the three most valuable and widely collected edible forest mushrooms: the American matsutake, morels, and chanterelles. The other species in Table 3.2 also deserve attention. The Forest Service and Bureau of Land Management sponsor many research projects in conjunction with the Pacific Northwest Research Station, but significant research is also being conducted by state universities and mycological societies. Collaboration among public land management agencies, private and corporate forest owners, research institutions, nongovernment organisations, researchers, managers, harvesters, and other stakeholders has greatly broadened the scope and enhanced the quality of past and present research projects. A summary of these research projects can be found in Pilz, Molina & Amaranthus (in press).

Many of the research projects being conducted in the Pacific Northwest are still active or not yet published. Our discussion, therefore, provides a cursory overview of emerging information as it applies to forest management and, where possible, citations for further information.

One of the first and most frequently asked questions about harvest sustainability was whether the greatly increased levels of harvesting would reduce subsequent levels of fruiting. Two approaches have been used to address this issue: experiments to examine harvesting impacts and monitoring projects to detect trends in productivity. Research conducted by the Oregon Mycological Society (Norvell, 1995) concurs with that found by Egli, Ayer & Chatelain (1990) in Switzerland, that picking mushrooms *per se* does not diminish subsequent fruiting over the period of a decade or more. Search methods, however, can have some impact. Recent studies (D. Pilz and others, unpublished data) of raking forest litter to find young matsutake suggest that raking into the mycelial layer can interfere with fruiting for several years, but that recovery is hastened by replacing the duff. These studies, however, do not address the long-term and broad-scale impacts of intensive commercial harvesting, forest management activities, or regional threats to forest health. Assessing these factors will require long-term monitoring and silvicultural research.

Several research projects in the Pacific Northwest have obtained baseline data on edible mushroom productivity: specifically, the seasonal (summed within a fruiting season) productivity per unit area (by using systematic sampling designs) of given forest stands or habitat types. These productivity values, derived from biologically diverse mixed-conifer stands, tend to be lower than edible mushroom productivity estimates reported elsewhere in the literature (Ohenoja & Koistinen, 1984; Ohenoja, 1988; Slee, 1991; Kalamees & Silver, 1993). Values range from less than 1 kg to as high as 20 kg per hectare per year for chanterelles and American matsutake (Pilz, Molina & Liegel, 1998; Pilz *et al.*, 1999). Typical values for commercially harvested stands range from 3 kg to 10 kg per hectare per year (Pilz & Molina, in press). A common goal of productivity studies conducted by the PNW Research Station is the development of statistically valid, practical and cost-effective monitoring protocols for edible mushrooms (Arnolds, 1992; Pilz & Molina, in press; Vogt *et al.*, 1992).

Sustaining harvests of edible mushrooms that are ectomycorrhizal depends on understanding the relations between the silvicultural characteristics of a stand and mushroom productivity. Tree species, age, growth rates, and density can all potentially influence the quantity of carbohydrates available to ectomycorrhizal fungi for growth and fruiting. Understanding these relations could lead to predictive models that would allow silviculturalists to ascertain the consequences of their stand prescriptions for mushroom productivity. The influence of thinning young stands on edible mushroom productivity is of particular interest in the Pacific Northwest because many forest tree plantations will be commercially thinned in the coming decades, especially on Federal lands where mushroom harvesting is allowed. Preliminary evidence indicates that heavy thinning (many trees removed) can substantially reduce mushroom productivity immediately after the harvest, but that lighter thinning (fewer trees removed) has a lesser impact on productivity (D. Pilz & R. Molina, unpublished data). The rate at which mushroom productivity rebounds as the remaining trees reoccupy a site has yet to be determined. Mushroom productivity might rebound to higher levels as the remaining trees grow more vigorously, but the complicating influence of soil compaction from ground-based thinning equipment will need to be factored into the interpretation of results.

Several genetic studies of edible mushrooms are showing greater species diversity than previously documented by alpha taxonomy. For instance, the west coast Pacific golden chanterelle, previously considered the same as the European *Cantharellus cibarius* Fr. has been shown actually to consist of three or more species, none exactly the same as *C. cibarius* (Redhead *et al.*, 1997; Dunham *et al.*, 1998). Similarly, morel studies in eastern Oregon have revealed five species of morels fruiting on a burned site, some not yet described (D. Pilz and others, unpublished data). Several mushroom species are harvested as 'king boletes' in the Pacific Northwest (*Boletus aereus* Fr., *B. barrowsii* Smith, and *B. edulis*), and *Boletus edulis* itself might actually comprise a species complex here. This level of species diversity is not surprising given the diversity of climates, habitats, soils, and ectomycorrhizal host tree species found in the region. The diversity does, however, present additional challenges to forest managers because unique ecological adaptations likely exist for each species, and they could respond differently to forest management choices. These are just a few examples of the management implications of recent edible mushroom research. Additional research and long-term monitoring are needed, however, to ensure the sustainability of widespread, intensive commercial mushroom harvesting and healthy populations of edible forest fungi.

Research and monitoring for sustainability

Most of the abundant and commercially harvested forest mushrooms, with the exception of some morels, are ectomycorrhizal species. The plentiful carbohydrate nutrition they derive from forest trees probably contributes to their abundant fruit body production. Their mycelial colonies are long-lived, and harvesting does not seem to impair fruiting in the short term. Indeed, humans world-wide have harvested edible forest mushrooms since ancient times (Arora, 1999; and see Chapter 8). So, a perfectly reasonable question is, 'Why bother monitoring their productivity in the first place?'

Managers of Federal lands in the United States are required to monitor the commercial harvest of natural resources to determine sustainable harvest levels and avoid degrading the environment. In addition to this legal mandate, some of the factors that have contributed to declines of edible mushrooms in Europe (Arnolds, 1991) and Japan (Hosford *et al.*, 1997) are also beginning to impact forests in the Pacific Northwest. These factors include pollution from growing urban areas, thinning ozone, climate change, introduced pathogens, and intensive timber management. Lastly, in any system that has functional thresholds, changes in quantity can translate into changes in kind when the thresholds are exceeded. Commercially harvesting edible forest mushrooms for international markets arguably entails greater and more widespread harvesting pressure than existed historically, and changes such as reduced (or enhanced) spore dispersal or trampling could have unforeseen long-term consequences.

The effort and resources applied to monitoring should be commensurate with resource value, socially acceptable levels of risk, current understanding of how resilient a renewable resource is to harvesting, and perceived threats to the resource. Pilz & Molina (1998, 2000) have proposed a regional programme that combines short-term silvicultural research with long-term but modest monitoring activities. The proposed programme has three components. The research component is intended to develop predictive models relating silvicultural activities to mushroom productivity. Ideally, a widely applicable model can be developed from data on several mushroom species and a range of sites and forest types.

The two monitoring components entail commercially harvested sites and sites where neither mushroom nor timber harvesting are allowed. On the commercially harvested sites, landowners would give co-operating harvesters exclusive access to discrete areas in exchange for information on what is harvested each year. On the natural sites, landowners could co-operate with mycological societies to estimate productivity without harvesting the sampled mushrooms. With various landowners, sites, and co-operators scattered around the region, long-term monitoring of potential trends in fruiting could be accomplished with a minimum of investment by any one participating agency or organisation. The use of standardised sampling methods, prearranged analyses, and a web site for posting annual results would provide all participants with timely results and regular encouragement to remain involved. The flexible co-operative organisation of the monitoring programme would allow co-operators to join or leave the programme as their interest and budgets dictate, although the best information will come from sites that are monitored for many years.

Regardless of further developments in research and monitoring in the Pacific Northwest, commercial harvesting of edible forest mushrooms likely will continue. The resource promises to be renewable and sustainable as long as we maintain healthy forests, understand the impacts of our actions, and remain alert to possible trends.

Concluding remarks

As noted throughout this chapter, our knowledge is sparse regarding the ecology and natural history of forest macrofungi. There are especially large gaps in our understanding of habitat requirements needed by species for maintaining healthy populations within changing landscapes. Thus, the challenge of conserving this biologically rich array of species remains considerable. Fungal conservation issues in the Pacific Northwest as outlined are unprecedented in scale and scope, and so our research programme includes a broad suite of ecological topics that provide a broad knowledge base to apply mycological principles to forest management plans. We use an ecosystem approach to plan and conduct our research at different scales because it provides opportunities to work closely with other forestry and ecological disciplines so that results can be woven into integrated solutions for broad-scale forest management. Fungi are critical ecosystem components, and their immense biological and functional diversity perform essential processes that lead to ecosystem resiliency and health. Our ability to demonstrate and communicate this primary fungal characteristic to other biological scientists, land managers, and the public is key to promoting fungal conservation issues.

We offer the following set of summary statements for mycologists to consider when planning and conducting conservation efforts for fungi.

- Set clear objectives for the study or inventory project. Knowing how the information will be used, and by whom, will guide the assemblage and proper communication of the findings.
- Select the appropriate methodology to meet the objectives. Data should be collected at the same scale at which the organism or process of interest is operating.
- Address mycological concerns within the context of local and regional land and natural resource management issues. This allows for fungi to be considered as part of the land management planning process.
- Explore opportunities to work with research and management colleagues across science disciplines. When possible, conduct mycological investigations within integrated studies that include other taxa and shared objectives such as overall maintenance of biodiversity.
- Emphasise the characterisation of habitat for mycological studies. Protecting fungal habitat at various scales of space and time will provide the most likely solution to the long-term conservation of fungi.
- Develop monitoring projects to inventory fungi periodically at long-term study sites to detect trends in populations. Enlist the help of local mycological societies in data collection.
- Develop quantitative models to help managers make decisions.
- Publish literature and conduct workshops to educate land managers and the public continuously about the importance of

fungi in maintaining healthy ecosystems. They are critical allies and partners in helping to set the conservation agenda for fungi.

References

Agee, J. K. (1993). *Fire Ecology of Pacific Northwest Forests.* Island Press: Washington, DC.

Agerer, R. (ed.) (1987–1998). *Colour Atlas of Ectomycorrhizas.* 1st–11th delivery. Einhorn-Verlag: Schwäbisch Gmünd, Germany.

Amaranthus, M. P., Parrish, D. S. & Perry, D. A. (1989). Decaying logs as moisture reservoirs after drought and wildfire. In *Stewardship of Soil, Air and Water Resources*, Proceedings of Watershed 89 (ed. E. Alexander), pp. 191–194. USDA Forest Service. R10–MB-77. USDA Forest Service, Alaska Region: Juneau, Alaska.

Amaranthus, M., Trappe, J. M., Bednar, L. & Arthur, D. (1994). Hypogeous fungal production in mature Douglas-fir forest fragments and surrounding plantations and its relation to coarse woody debris and animal mycophagy. *Canadian Journal of Forest Research* **24**, 2157–2165.

Ammirati, J. (1979). Chemical studies of mushrooms: the need for voucher specimens. *Mycologia* **71**, 437–441.

Arnolds, E. (1989). A preliminary RED data list of macrofungi in The Netherlands. *Persoonia* **14**, 77–125.

Arnolds, E. (1991). Decline of ectomycorrhizal fungi in Europe. *Agriculture, Ecosystems and Environment* **35**, 209–244.

Arnolds, E. (1992). Mapping and monitoring of macromycetes in relation to nature conservation. *McIlvainea* **10**, 4–27.

Arnolds, E. & De Vries, B. (1993). Conservation of fungi in Europe. In *Fungi in Europe: Investigation, Recording and Conservation* (ed. D. N. Pegler, L. Boddy, B. Ing & P. M. Kirk), pp. 211–230. Royal Botanic Gardens: Kew, UK.

Arora, D. (1999). The way of the wild mushroom. *California Wildlife* **52**, 8–19.

Avise, J. C. (1994). *Molecular Markers, Natural History, and Evolution.* Chapman and Hall: New York.

Baar, J., Horton, T. R., Kretzer, A. M. & Bruns, T. D. (1999). Mycorrhizal colonization of *Pinus muricata* from resistant propagules after a stand-replacing wildfire. *New Phytologist* **143**, 409–418.

Baar, J., Ozinga, W. A. & Kuyper, T. H. W. (1994). Spatial distribution of *Laccaria bicolor* genets reflected by sporocarps after removal of litter and humus layers in a *Pinus sylvestris* forest. *Mycological Research* **98**, 726–728.

Bas, C. (1978). Veranderingen in de Nederlandse paddestolenflora. *Coolia* **21**, 98–104.

Bormann, B. T., Cunningham, P. G., Brookes, M. H., Manning, V. M. & Collopy, M. W. (1994). *Adaptive Ecosystem Management in the Pacific Northwest.* General Technical Report PNW-GTR-341. US Department of Agriculture, Forest Service, Pacific Northwest Research Station: Portland, Oregon.

Bossart, J. L. & Prowell, D. P. (1998). Genetic estimates of population structure and gene flow: Limitations, lessons, and new directions. *Trends in Ecology and Evolution* **13**, 202–206.

Brown, D., Goetz, C., Rogers, M. & Westfall, D. (1985). *Mushrooming in the Pacific Northwest. NAMA XXV.* [25th Anniversary Issue], pp. 29–35. The North American Mycological Association: Charleston, West Virginia.

Brownlee, C. J., Duddridge, A., Malibari, A. & Read, D. J. (1983). The structure and function of mycelial systems of ectomycorrhizal roots with special reference to their role in forming inter-plant connections and providing pathways for assimilate and water transport. *Plant and Soil* **71**, 433–443.

Bruns, T. D., Fogel, R., White, T. J. & Palmer, J. D. (1989). Accelerated evolution of a false-truffle from a mushroom ancestor. *Nature* **339**, 140–142.

Bruns, T. D., Szaro, T. M., Gardes, M., Cullings, K. W., Pan, J. J., Taylor, D. L., Horton, T. R., Kretzer, A., Garbelotto, M. & Li, Y. (1998). A sequence database for the identification of ectomycorrhizal basidiomycetes by phylogenetic analysis. *Molecular Ecology* **7**, 257–272.

Carey, A. B., Thysell, D. R. & Brodie, A. W. (1999). *The Forest Ecosystem Study: Background, Rational, Implementation, Baseline Conditions, and Silvicultural Assessment*. USDA Forest Service General Technical Report, PNW-GTR-457: Portland, Oregon.

Castellano, M. A. (1997*a*). Recent developments concerned with monitoring and inventory of fungal diversity in northwestern United States. In *Mycology in Sustainable Development: Expanding Concepts, Vanishing Borders* (ed. M. Palm & I. Chapela), pp. 93–98. Parkway Publishing Inc.: Boone, North Carolina.

Castellano, M. A. (1997*b*). Towards a RED list for Oregon fungi. In *Conservation and Management of Native Plants and Fungi* (ed. T. N. Kaye, A. Liston, R. M. Love, D. L. Luoma, R. J. Meinke & M. V. Wilson), pp. 222–226. Native Plant Society of Oregon: Corvallis, Oregon.

Castellano, M. A. (1998). Toward a RED list for Idaho's macrofungi. In *Highlighting Idaho's Rare Fungi and Lichens*, Management Technical Bulletin 98–1 (ed. R. Rosentreter & A. DeBolt). Idaho Bureau of Land Management: Idaho.

Castellano, M. A. & O'Dell, T. (1997). *Management Recommendations for Survey and Manage Fungi.* USDA Forest Service, Regional Ecosystem Office: Portland, Oregon.

Castellano, M. A., Smith, J. E., O'Dell, T. E., Cazares, E. & Nugent, S. (1999). *Handbook to Strategy 1 Fungal Species in the Northwest Forest Plan*. USDA Forest Service General Technical Report PNW-GTR-476. US Department of Agriculture, Forest Service, Pacific Northwest Research Station: Portland, Oregon.

Ciesla, W. M. (1998). *Non-wood Forest Products from Conifers. Non-wood Forest Products 12*. FAO, United Nations: Rome.

Colgan, W. (1997). *Diversity, Productivity, and Mycophagy of Hypogeous Mycorrhizal Fungi in a Variable Thinned Douglas-fir Forest.* Ph. D. Thesis, Oregon State University.

Colgan, W., Carey, A. B., Trappe, J. M., Molina, R. & Thysell, D. (1999). Diversity and productivity of hypogeous fungal sporocarps in a variably thinned Douglas-fir forest. *Canadian Journal of Forest Research* **29**, 1259–1268.

Dahlberg, A., Jonsson, L. & Nylund, J.-E. (1997). Species diversity and distribution of biomass above and below ground among ectomycorrhizal fungi in an old Norway spruce forest in South Sweden. *Canadian Journal of Botany* **75**, 1323–1335.

Dahlberg, A. & Stenlid, A. J. (1990). Population structure and dynamics in *Suillus bovinus* as indicated by spatial distribution of fungal clones. *New Phytologist* **115**, 487–493.

Danielson, R. M. (1984). Ectomycorrhiza association in jack pine stands in northeastern Alberta. *Canadian Journal of Botany* **62**, 932–939.

Dighton, J. & Mason, P. A. (1985). Mycorrhizal dynamics during forest tree development. In *Developmental Biology of Higher Fungi* (ed. D. Moore, L. A. Casselton, D. A. Wood & J. C. Frankland), pp. 117–139. Cambridge University Press: Cambridge, UK.

Dizon, A. E., Taylor, B. L. & O'Corry-Crowe, G. M. (1995). Why statistical power is necessary to link analyses of molecular variation to decisions about population structure. In *Evolution and the Aquatic Ecosystem: Defining Unique Units in Population Conservation* (ed. J. L. Nielsen & D. A. Powers). American Fisheries Society: Bethesda, Maryland.

Dreisbach, T. A., Smith, J. E. & Molina, R. (in press). Challenges of modeling fungal habitat: when and where do you find chanterelles? In *Predicting Species Occurrence: Issues of Scale and Accuracy* (ed. J. M. Scott, P. J. Heglund, M. Morrision, M. Raphael, J. Haufler & B. Wall). Island Press: Covello, California.

Dunham, S., O'Dell, T., Molina, R. & Pilz, D. (1998). Fine scale genetic structure of chanterelle (*Cantharellus formosus*) patches in forest stands with different disturbance intensities. [Abstract] In *Programme and Abstracts, Second International Conference on Mycorrhiza*, July 5–10, 1998 (ed. U. Ahonen-Jonnarth, E. Danell, P. Fransson *et al.*). Swedish University of Agricultural Sciences: Uppsala, Sweden.

Egli, S., Ayer, F. & Chatelain, F. (1990). Der Einfluss des Pilzsammelns auf die Pilzflora. *Mycologia Helvetica* **3**, 417–428.

Epperson, B. K. (1993). Recent advances in correlation studies of spatial patterns of genetic variation. *Evolutionary Biology* **27**, 95–155.

FEMAT (Forest Ecosystem Management Assessment Team) (1993). *Forest Ecosystem Management: An Ecological, Economic and Social Assessment* [irregular pagination]. US Department of Agriculture, US Department of the Interior (and others): Portland, Oregon.

Finlay, R. D. & Read, D. J. (1986). The structure and function of the vegetative mycelium of ectomycorrhizal plants. I. Translocation of ^{14}C-labelled carbon between plants interconnected by a common mycelium. *New Phytologist* **103**, 143–156.

Gardes, M. & Bruns, T. D. (1993). ITS primers with enhanced specificity for basidiomycetes – application to the identification of mycorrhizae and rusts. *Molecular Ecology* **2**, 113–118.

Gardes, M. & Bruns, T. D. (1996). Community structure of ectomycorrhizal fungi in a *Pinus muricata* forest: above- and below-ground views. *Canadian Journal of Botany* **74**, 1572–1583.

Gardes, M., White, T. J., Fortin, J. A., Bruns, T. D. & Taylor, J. W. (1991). Identification of indigenous and introduced symbiotic fungi in ectomycorrhizae by amplification of nuclear and mitochondrial ribosomal DNA. *Canadian Journal of Botany* **69**, 180–190.

Gehring, C. A., Theimer, T. C., Whitham ,T. G. & Keim, P. (1998). Ectomycorrhizal fungal community structure of pinyon pines growing in two environmental extremes. *Ecology* **79**, 1562–1572.

Goodman, D. M., Durall, D. M., Trofymow, J. A. & Berch, S. M. (1996). *A*

Manual of Concise Descriptions of North American Ectomycorrhizae. Mycologue Publications: Victoria, B.C., Canada.

Gryta, H., Debaud, J.-C., Effosse, A., Gay, G. & Marmeisse, R. (1997). Fine-scale structure of populations of the ectomycorrhizal fungus *Hebeloma cylindrosporum* in coastal sand dune forest ecosystems. *Molecular Ecology* **6**, 353–364.

Harmon, M. E. & Sexton, J. (1995). Water balance of conifer logs in early stages of decomposition. *Plant and Soil* **172**, 141–152.

Harvey, A. E., Jurgensen, M. F. & Larsen, M. J. (1978). Seasonal distribution of ectomycorrhizae in a mature Douglas-fir/larch forest soil in western Montana. *Forest Science* **24**, 203–208.

Harvey, A. E., Larsen, M. J. & Jurgensen, M. F. (1976). Distribution of ectomycorrhizae in a mature Douglas-fir/larch forest soil in western Montana. *Forest Science* **22**, 393–398.

Harvey, A. E., Larsen, M. J. & Jurgensen, M. F. (1980). Partial cut harvesting and ectomycorrhizae: early effects in Douglas-fir and larch of western Montana. *Canadian Journal of Forest Research* **10**, 436–440.

Hawksworth, D. L. (1991). The fungal dimension of biodiversity: its magnitude and significance. *Mycological Research* **95**, 441–456.

Horton, T. & Bruns, T. D. (1998). Multiple-host fungi are the most frequent and abundant ectomycorrhizal types in a mixed stand of Douglas fir (*Pseudotsuga menziesii*) and bishop pine (*Pinus muricata*). *New Phytologist* **139**, 331–339.

Horton, T., Cazares, E. & Bruns, T. D. (1998). Ectomycorrhizal, vesicular-arbuscular and dark septate fungal colonization of bishop pine (*Pinus muricata*) seedlings in the first 5 months after growth after wildfire. *Mycorrhiza* **8**, 11–18.

Hosford, D., Pilz, D., Molina, R. & Amaranthus, M. (1997). *Ecology and Management of the Commercially Harvested American Matsutake Mushroom*. General Technical Report. PNW-GTR-412. US Department of Agriculture, Forest Service, Pacific Northwest Research Station: Portland, Oregon.

Hughes, C. R. & Queller, D. C. (1993). Detection of highly polymorphic microsatellite loci in a species with little allozyme polymorphism. *Molecular Ecology* **2**, 131–137.

Hughes, K. W., McGhee, L. L., Methven, A. S., Johnson, J. E. & Petersen, R. H. (1999). Patterns of geographic speciation in the genus *Flammulina* based on sequences of the ribosomal ITS1–5.8S-ITS2 area. *Mycologia* **91**, 978–986.

Ingleby, K., Mason, P. A., Last, F. T. & Fleming, L. V. (1990). *Identification of Ectomycorrhizas*. Publication number 5, Institute of Terrestrial Ecology Research: London.

Jensen, M. E. & Everett, R. (1994). An overview of ecosystem management principles. In *Volume II: Ecosystem Management: Principles and Applications*, General Technical Report PNW-GTR-318 (ed. M. E. Jensen & P. S. Bourgeron), pp. 6–15. US Department of Agriculture, Forest Service, Pacific Northwest Research Station: Portland, Oregon.

Jonsson, L., Dahlberg, A., Nilsson, M.-C., Zackrisson, O. & Kårén, O. (1999). Ectomycorrhizal fungal communities in late-successional Swedish boreal forests, and their composition following wildfire. *Molecular Ecology* **8**, 205–215.

Kalamees, K. & Silver, S. (1993). Productivity of edible fungi in pine forests of the *Mytrillus* and *Vaccinium uliginosum* site types in North-West Estonia. *Aquilo Serie Botanica* **31**, 123–126.

Kårén, O., Högberg, N., Dahlberg, A., Jonsson, L. & Nylund, J.-E. (1997). Inter- and intraspecific variation in the ITS region of rDNA of ectomycorrhizal fungi in Fennoscandia as detected by endonuclease analysis. *New Phytologist* **136**, 313–325.

Kiester, A. R., Scott, J. M., Csuti, B., Noss, R. F., Butterfield, B., Sahr, K. & White, D. (1996). Conservation prioritization using GAP data. *Conservation Biology* **10**, 1332–1342.

Kropp, B. R. (1982). Rotten wood as mycorrhizal inoculum for containerized western hemlock. *Canadian Journal of Forest Research* **12**, 428–431.

Kuhnlein, H. V. & Turner, N. J. (1991). Traditional plant foods of Canadian indigenous peoples: nutrition, botany, and use. In *Food and Nutrition in History and Anthropology, Volume 8* (ed. S. H. Katz). Gordon and Breach: New York.

Largent, D. L. & Sime, A. D. (1995). A preliminary report on the phenology, sporulation, and lifespan of *Cantharellus cibarius* and *Boletus edulis* basidiomes in Patrick Point State Park. In *Symposium Proceedings, 43rd Annual Meeting of the California Forest Pest Council*, Nov. 16–17, 1994 (ed. D. H. Adams, J. E. Rios & A. J. Stoere). Rancho Cordova, California.

Love, T., Jones, E. & Liegel, L. (1998). Valuing the temperate rainforest: wild mushrooming on the Olympic Peninsula Biosphere Reserve. In *The Biological, Socioeconomic, and Managerial Aspects of Chanterelle Mushroom Harvesting: the Olympic Peninsula, Washington State,* USA, Ambio Special Report 9 (ed. L. H. Liegel), pp. 16–25. Royal Swedish Academy of Sciences: Stockholm, Sweden.

Lund, G., Pajari, B. & Korhonen, M. (eds) (1998). *Sustainable Development of Non-wood Goods and Benefits From Boreal and Cold Temperate Forests. Proceedings of the International Workshop. 18–22 January 1998, Proceedings No. 23*. European Forest Institute: Joensuu, Finland.

Luoma, D. L., Eberhart, J. L. & Amaranthus, M. P. (1996). Response of ectomycorrhizal fungi to forest management treatments – sporocarp production. In *Mycorrhizas in Integrated Systems: From Genes to Plant Development* (ed. C. Azcon-Aguilar & J. M. Barea), pp. 553–556. Office for Official Publications of the European Communities, European Commission: Luxembourg.

Magnussen, S. (1993). Bias in genetic variance estimates due to spatial autocorrelation. *Theoretical Applied Genetics* **86**, 349–355.

Maser, C., Trappe, J. M. & Nussbaum, R. A. (1978). Fungal–small mammal interrelationships with emphasis on Oregon coniferous forests. *Ecology* **59**, 799–809.

Maser, Z., Maser, C. & Trappe, J. (1985). Food habits of the northern flying squirrel (*Glaucomys sabrinus*) in Oregon. *Canadian Journal of Zoology* **63**, 1084–1088.

Miller, S. L., McClean, T. M., Stanton, N. L. & Williams, S. E. (1998). Mycorrhization, physiognomy, and first-year survivability of conifer seedlings following natural fire in Grand Teton National Park. *Canadian Journal of Forest Research* **28**, 115–122.

Moldenke, A. R. (1999). Soil-dwelling arthropods: their diversity and functional roles. In *Proceedings: Pacific Northwest Forest and Rangeland Soil Organism Symposium*, Corvallis, Oregon, March 17–19, 1998 (ed. R. T. Meurisse, W.

G. Ypsilantis & C. Seybold), pp. 33–44. General Technical Report PNW-GTR-461. US Department of Agriculture, Forest Service, Pacific Northwest Research Station: Portland, Oregon.

Molina, R. & Trappe, J. M. (1982). Patterns of ectomycorrhizal host specificity and potential among Pacific Northwest conifers and fungi. *Forest Science* **28**, 423–458.

Molina, R., Massicotte, H. B. & Trappe, J. M. (1992). Specificity phenomena in mycorrhizal symbioses: community-ecological consequences and practical implications. In *Mycorrhizal Functioning: An Integrative Plant–Fungal Process* (ed. M. F. Allen), pp. 357–423. Chapman and Hall: New York.

Molina, R., O'Dell, T., Luoma, D., Amaranthus, M., Castellano, M. & Russell, K. (1993). *Biology, Ecology, and Social Aspects of Wild Edible Mushrooms in the Forests of the Pacific Northwest: A Preface to Managing Commercial Harvest*. General Technical Report PNW-GTR-309. US Department of Agriculture, Forest Service, Pacific Northwest Research Station: Portland, Oregon.

North, M., Trappe, J. & Franklin, J. (1997). Standing crop and animal consumption of fungal sporocarps in Pacific Northwest forests. *Ecology* **78**, 1543–1554.

Norvell, L. (1995). Loving the chanterelle to death? The ten-year Oregon chanterelle project. *McIlvainea* **12**, 6–23.

O'Dell, T. E., Ammirati, J. F. & Schreiner, E. G. (1999). Species richness and abundance of ectomycorrhizal basidiomycete sporocarps on a moisture gradient in the *Tsuga heterophylla* zone. *Canadian Journal of Botany* **77**, 1699–1711.

Ohenoja, E. (1988). Effect of forest management procedures on fungal fruit body production in Finland. *Acta Botanica Fennica* **136**, 81–84.

Ohenoja, E. & Koistinen, R. (1984). Fruit body production of larger fungi in Finland. 2. Edible fungi in northern Finland, 1976–1978. *Annales Botanici Fennici* **21**, 357–366.

Parke, J. L., Linderman, R. G. & Trappe, J. M. (1984). Inoculum potential of ectomycorrhizal fungi in forest soils of southwest Oregon and northern California. *Forest Science* **30**, 300–304.

Patton, D. R. (1997). *Wildlife Habitat Relationships in Forested Ecosystems*. Timber Press: Portland, Oregon.

Perry, D. A., Amaranthus, M. P., Borchers, J. G., Borchers, S. L. & Brainerd, R. E. (1989). Bootstrapping in ecosystems. *BioScience* **39**, 230–237.

Pilz, D. & Molina, R. (eds) (1996). *Managing Forest Ecosystems to Conserve Fungus Diversity and Sustain Wild Mushroom Harvest.* General Technical Report PNW-GTR-371. US Department of Agriculture, Forest Service, Pacific Northwest Research Station: Portland, Oregon.

Pilz, D. & Molina, R. (1998). A proposal for regional monitoring of edible forest mushrooms. *Mushroom, The Journal of Wild Mushrooming* **3**, 19–23.

Pilz, D. & Molina, R. (in press). Sustaining productivity of commercially harvested edible forest mushrooms. *Forest Ecology and Management*.

Pilz, D. P. & Perry, D. A. (1984). Impact of clearcutting and slash burning on ectomycorrhizal associations of Douglas-fir seedlings. *Canadian Journal of Forest Research* **14**, 94–100.

Pilz, D., Molina, R. & Amaranthus, M. P. (in press). Productivity and sustainable harvest of edible forest mushrooms: current biological research and new directions in Federal monitoring. In *Nontimber Forest Products in*

the United States: An Overview of Research and Policy Issues (ed. M. R. Emery & R. J. McClain). Special Issue of *Journal of Sustainable Forestry*.
Pilz, D., Molina, R. & Liegel, L. (1998). Biological productivity of chanterelle mushrooms in and near the Olympic Peninsula Biosphere Reserve. In *The Biological, Socioeconomic, and Managerial Aspects of Chanterelle Mushroom Harvesting: The Olympic Peninsula, Washington State, USA* (ed. L. H. Liegel), pp. 8–13. Ambio Special Report 9. Royal Swedish Academy of Sciences: Stockholm, Sweden.
Pilz, D., Smith, J., Amaranthus, M. P., Alexander, S., Molina, R. & Luoma, D. (1999). Mushrooms and timber: managing commercial harvesting in the Oregon Cascades. *Journal of Forestry* **97**, 4–11.
Redhead, S. A., Norvell, L. L. & Danell, E. (1997). *Cantharellus formosus* and the Pacific Golden Chanterelle harvest in western North America. *Mycotaxon* **65**, 285–322.
Richardson, M. J. (1970). Studies on *Russula emetica* and other agarics in a scots pine plantation. *Transactions of the British Mycological Society* **55**, 217–229.
Ruckelshaus, M. H. (1998). Spatial scale of genetic structure and an indirect estimate of gene flow in eelgrass, *Zostera marina*. *Evolution* **52**, 330–343.
Schlosser, W. E. & Blatner, K. A. (1995). The wild edible mushroom industry of Washington, Oregon and Idaho. *Journal of Forestry* **93**, 31–36.
Schlumpf, E. (1976). Sollen unsere Pilze aussterben? *Schweizerische Zeitschrift Pilzkunde* **54**, 101–105.
Schneider, E. (1999). Favored fungi: part one. *Food Arts*, October, 158–167.
Schowalter, T., Hansen, E., Molina, R. & Zhang, Y. (1997). Integrating the ecological roles of phytophagous insects, plant pathogens, and mycorrhizae in managed forests. In *Creating A Forestry for the 21st Century; the Science of Ecosystem Management* (ed. K. A. Kohm & J. F. Franklin), pp. 171–189. Island Press: Washington, DC.
Simard, S. W., Perry, D. A., Jones, M. D., Myrold, D., Durral, D. M. & Molina, R. (1997). Net transfer of carbon between ectomycorrhizal tree species in the field. *Nature* **388**, 579–582.
Sime, A. D. & Petersen, R. H. (1999). Intercontinental interrelationships among disjunct populations of *Melanotus* (Strophariaceae, Agaricales). *Mycotaxon* **71**, 481–492.
Slatkin, M. & Barton, N. H. (1989). A comparison of three indirect methods for estimating average levels of gene flow. *Evolution* **4**, 1349–1368.
Slee, R. W. (1991). The potential of small woodlands in Britain for edible mushroom production. *Scottish Forestry* **45**, 3–12.
Smith, M. L., Bruhn, J. N. & Anderson, J. B. (1992). The fungus *Armillaria bulbosa* is among the largest and oldest living organisms. *Nature* **356**, 428–431.
Smith, S. E. & Read, D. J. (1997). *Mycorrhizal Symbiosis*. Academic Press: London.
Sokal, R. R. & Oden, N. L. (1991). Spatial autocorrelation analysis as an inferential tool in population genetics. *American Naturalist* **138**, 518–521.
Starfield, A. M. & Bleloch, A. L. (1991). *Building Models for Conservation and Wildlife Management*. Burgess International Group: Edina, Minnesota.
Stendall, E. R., Horton, T. R. & Bruns, T. D. (1999). Early effects of prescribed fire on the structure of the ectomycorrhizal fungus community in a Sierra Nevada ponderosa pine forest. *Mycological Research* **103**, 1353–1359.
Sullivan, R. (1998). The mushroom wars. *Men's Journal* (Sept.), 75–79 &

170–175.

Swift, M. J. (1982). Basidiomycetes as components of forest ecosystems. In *Decomposer Basidiomycetes: Their Biology and Ecology* (ed. J. C. Frankland, J. N. Hedger & M. J. Swift), pp. 307–338. Cambridge University Press: Cambridge, UK.

Taylor, D. L. & Bruns, T. D. (1999). Community structure of ectomycorrhizal fungi in a *Pinus muricata* forest: minimal overlap between the mature forest and resistant propagule communities. *Molecular Ecology* **8**, 1837–1850.

Termorshuizen, A. J. (1991). Succession of mycorrhizal fungi in stands of *Pinus sylvestris* in The Netherlands. *Journal of Vegetation Science* **2**, 555–564.

Thompson, S. K. (1992). *Sampling*. Wiley Interscience: New York.

Todd, N. K. & Rayner, A. D. M. (1980). Fungal individualism. *Science Progress (Oxford)* **66**, 331–354.

Trappe, J. M. (1977). Selection of fungi for ectomycorrhizal inoculation in nurseries. *Annual Review of Phytopathology* **15**, 203–222.

US Department of Agriculture, Forest Service; US Department of the Interior, Bureau of Land Management (1994*a*). *Final Supplemental Environmental Impact Statement on Management of Habitat for Late-successional and Old-growth Forest Related Species Within the Range of the Spotted Owl.* Vol. 1, US Department of Agriculture, Forest Service, US Department of the Interior, Bureau of Land Management: Washington, DC.

US Department of Agriculture, Forest Service; US Department of the Interior, Bureau of Land Management (1994*b*). *Record of Decision for Amendments to Forest Service and Bureau of Land Management Planning Documents Within the Range of the Northern Spotted Owl.* [Place of publication unknown].

US Department of Agriculture, Forest Service; US Department of the Interior, Bureau of Land Management (1996). *Integrated Scientific Assessment for Ecosystem Management in the Interior Columbia Basin and Portions of the Klamath and Great Basins.* General Technical Report PNW-GTR-392. US Department of Agriculture, Forest Service, Pacific Northwest Research Station: Portland, Oregon.

US Fish & Wildlife Service (1981). *Standards for the Development of Habitat Suitability Index Models.* 103–ESM. US Department of the Interior: Washington DC.

Visser, S. (1995). Ectomycorrhizal fungal succession in jack pine stands following wildfire. New *Phytologist* **129**, 389–401.

Vogt, K. A., Bloomfield, J., Ammirati, J. F. & Ammirati, S. R. (1992). Sporocarp production by Basidiomycetes, with emphasis on forest ecosystems. In *the Fungal Community: its Organization and Role in the Ecosystem*, Second edition (ed. G. C. Carroll & D. T. Wicklow), pp. 563–581. Marcel Dekker, Inc.: New York.

Washington Natural Heritage Program (1997). *Endangered, Threatened and Sensitive Vascular Plants of Washington – with Working Lists of Rare Non-vascular Plants.* Department of Natural Resources: Olympia, Washington.

Waters, J. R., McKelvey, K. S., Luoma, D. L. & Zabel, C. J. (1997). Truffle production in old-growth and mature fir stands in northeastern California. *Forest Ecology and Management* **96**, 155–166.

Waters, J. R., McKelvey, K. S., Zabel, C. J. & Oliver, W. W. (1994). The effects of thinning and broadcast burning on sporocarp production of hypogeous fungi. *Canadian Journal of Forest Research* **24**, 1516–1522.

White, T. J., Bruns, T., Lee, S. & Taylor, J. (1990). Amplification and direct sequencing of fungal ribosomal RNA genes for phylogenetics. In *PCR Protocols: A Guide to Methods and Applications* (ed. M. A. Innis, D. H. Gelfand, J. J. Sninsky & T. J. White), pp. 315–322. Academic Press: San Diego, California.

Wiens, J. A. (1989). Spatial scaling in ecology. *Functional Ecology* **3**, 385–397.

Wiens, J. A. (1997). Metapopulation dynamics and landscape ecology. In *Metapopulation Biology: Ecology, Genetics, and Evolution* (ed. I. A. Hanski & M. E. Gilpin). Academic Press: London.

Wilkins, W. H. & Harris, G. C. M. (1946). The ecology of the larger fungi. V. An investigation into the influence of rainfall and temperature on the seasonal production of fungi in a beechwood and a pinewood. *Annals of Applied Biology* **33**, 179–188.

Wood, M. (2000). North American mycological societies. MykoWeb. http://www.mykoweb.com/index.html.

Wright, S. (1951). The genetical structure of populations. *Annual Eugenetics* **15**, 323–354.

Wright, S. (1969). *Evolution and Genetics of Populations. Vol. 2. The Theory of Gene Frequencies*. University of Chicago Press: Chicago, Illinois.

Zak, B. (1973). Classification of ectomycorrhizae. In *Ectomycorrhizae: Their Ecology and Physiology* (ed. G. C. Marks & T. T. Kozlowski), pp. 43–78. Academic Press: New York.

4

The future of fungi in Europe: threats, conservation and management

EEF ARNOLDS

Introduction

During the 9th Congress of European Mycologists in Oslo, 1985, quantitative data were presented for the first time on changes in the mycota, in particular in The Netherlands (Arnolds, 1988). These data supported the fears of many mycologists and naturalists, namely that macrofungi were decreasing, at least some species in some areas. This information was the motive for the establishment of the European Committee for Protection of Fungi, which later changed its name into the European Council for Conservation of Fungi (ECCF). In this chapter I will review the developments in fungus conservation in Europe during the past 15 years and look forward towards challenges in the future. I restrict myself to macrofungi since microfungi are rarely considered in relation to conservation (but see Chapter 17).

Effective conservation of fungi depends on the compilation and integration of data from different disciplines, here summarised in the scheme shown in Fig. 4.1. Central in this scheme are Red Data lists because they reflect our knowledge on the status of fungal species and are meant as a basis for measurements and to present our conclusions to a wider audience. Obviously, progress in conservation depends on progress in other mycological disciplines, such as taxonomy, distribution and ecology. Therefore I shall pay some attention to these disciplines as well.

European Council for Conservation of Fungi

The ECCF has played a central role in further development of methods to study changes in the mycota, in stimulating the compilation of Red Data lists and in spreading the message that fungi are in need of conservation

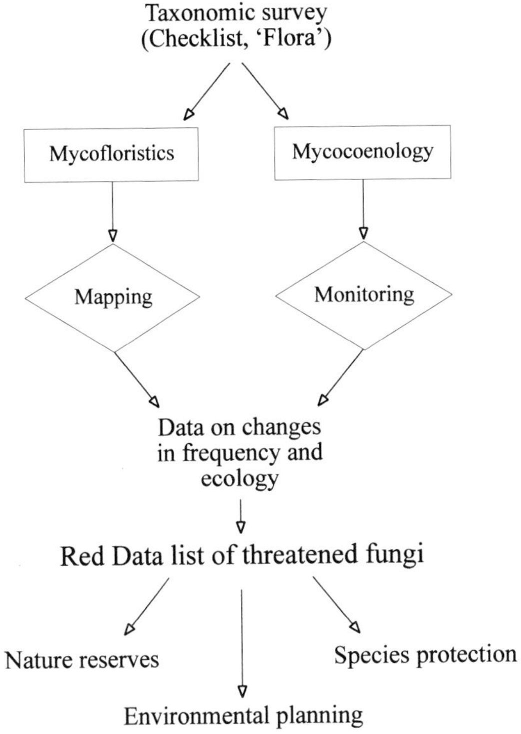

Fig. 4.1. Schematic representation of relations between conservation of fungi and other disciplines of mycology.

measures, like many other organisms. It is the international informal forum for exchange of ideas, and particularly valuable ECCF meetings were those in Lódz (1988, proceedings edited by Jansen & Ławrynowicz, 1991), Vilm, Germany (1991, proceedings edited by Arnolds & Kreisel, 1993), Le Louvrin, Switzerland (1993) and Vipeteno, Italy (1997, proceedings edited by Perini, 1998). Official representatives attended these meetings as well as other people involved in conservation from many European countries. In between these meetings, business meetings of the ECCF were held at intermittent Congresses of European Mycologists. Since the Oslo congress, conservation has become a constant and substantial topic at European congresses. For instance, one third of the papers in the proceedings of the Kew congress were devoted to conservation or related subjects (Pegler *et al.*, 1993).

The consideration of conservation aspects was stimulated by several factors, among which were:

- Mycologists and ecologists were concerned about the possible decline of fungi and the likely consequences for the functioning of ecosystems.
- Knowledge of ecology and geographical distribution of fungi has strongly increased in some countries, which enabled an analysis of trends of various species.
- Public and political interest in nature conservation has strongly increased, culminating in international treaties on biodiversity and protection of endangered organisms (e.g. Bern convention).

Taxonomy

Conservation depends on good taxonomic knowledge since species are the main units to be dealt with. Much progress has been made in recent years in taxonomy of European fungi, both by thorough classical revisions based on morphological characters and by the application of new methods, such as molecular techniques and compatibility tests. Despite all the recent effort, the work is not yet finished. Numerous new species are described each year and many critical groups are awaiting modern revisions.

A practical problem in conservation is the existence of different morphological species concepts in different parts of the world. In some regions relatively broad concepts are prevailing, for example in Scandinavia and Germany, in others relatively narrow concepts are favoured, for example in France by taxonomists like Bon, Henry and others. I mention only one out of many examples: the blackening taxa of *Hygrocybe* are described by Boertmann (1995) as one species with two varieties and by Courtecuisse & Duhem (1994) as seven distinct species.

In some cases the application of modern techniques can support a particular point of view, but in other cases the problems are becoming rather more complicated. For instance, Aanen & Kuyper (1999) found in the group of *Hebeloma crustuliniforme* twenty compatibility groups, representing biological species, but they were able to recognise at most four species on the basis of morphological characters. Biological species are impractical units in field mycology when it is impossible to identify them by morphological characters.

Divergence of species concepts is a major problem when comparing species diversity, checklists and Red Data lists of different countries. It means that mycologists do not speak a common language and this may produce irritation in the nonmycological world. This problem is increased further by substantial differences in nomenclature. In my opinion it is a challenge to European taxonomists to compile a common European checklist with recommended names. Such a survey should be compiled on the basis of national checklists, which are also quite useful for the compilation of national Red Data lists. Dennis *et al.* (1960) in Great Britain published the first modern checklist. It is a taxonomic list with references to synonyms, plates and descriptions. A variant with more relevant information for conservationists is the ecological checklist of Sweden, providing concise information on habitat and distribution patterns (Hallingbäck, 1994). An extended combination of a taxonomic and ecological checklist appeared in The Netherlands (Arnolds *et al.*, 1995). It is the basic document for mapping, the national Red Data list and conservation in that country. The list of The Netherlands contains almost 3500 species, including 2700 basidiomycetes. A recent checklist of basidiomycetes in Greece enumerates only 811 species (Zervakis *et al.*, 1998). Apparently the mycota of Greece is not yet fully explored, as the authors also point out. Nevertheless, such a list is a major achievement as a basis for further exploration of the regional mycota and as a contribution to knowledge at the European level. Any checklist is better than no checklist.

Distribution and mapping

Reliable data on present and former distributions, as well as frequency of fungi, are indispensable for compilation of Red Data lists. The actual distribution is a clue to the rarity of species and their degree of endemism. Data from the past are needed to establish possible increase or decrease. Before 1970 no systematic recording of fungi took place in Europe. This is a serious problem in drawing conclusions on declining species and therefore in assigning Red List categories to fungi on an objective basis.

In the 1960s a European mapping programme was initiated for 100 selected species. The results were interesting but not really representative because of methodological problems and uneven recording in various parts of Europe. In many cases the maps reflected the distribution of mycologists rather than of fungi (Lange, 1974). Nevertheless this attempt had a stimulatory effect. Mapping programmes have been initiated since 1970 in several European countries. The methodology differs considerably from one country to another, for instance concerning grid units, collected data and format of the database. Up to now, complete distribution atlases have been published for Germany (Krieglsteiner, 1991, 1993) and The Netherlands (Anon., 2000), without comments on the patterns of the species. In addition, in The Netherlands an atlas was published treating 370 species extensively, with data on changes in frequency, habitat and periodicity (Nauta & Vellinga, 1995). The latter type of atlas is, of course, a much more valuable source of information in relation to conservation. In many other countries maps have been published of selections of species, for instance of threatened polypores in Finland (Kotiranta & Niemelä, 1996). Mapping was carried out until recently mainly in northwestern Europe, but at present projects have also been launched in Greece and Ukraine, for instance.

National mapping programmes are the essential bricks for construction of any European survey. It is sometimes felt to be a great problem that the databases are not directly compatible because of, for instance, differences in scale of map grids. In my opinion this is a technical problem that can be solved by modern database management. It is inevitable that different methods are developed, depending on national goals and possibilities. The participation of well-trained amateur field mycologists in mapping programmes is extremely useful, as the results in The Netherlands and Germany demonstrate.

In 1999 the ECCF launched a new pilot project for producing European maps of 20 selected, threatened species, based on a compilation of existing data. Such projects are essential to bring conservation of fungi to the attention of European authorities and may, moreover, stimulate interest in national mapping. Ideally, all distribution data should be included in a European database, but such an initiative depends on availability of researchers and funding.

Ecology

Ecological knowledge of fungi is necessary for identification of valuable habitats, conservation of these habitats and the realisation of proper management. Data on autecology of species mainly come from compilation of field surveys and are summarised in some checklists (Arnolds *et al.*, 1995; Hallingbäck, 1994). Mycocoenological studies have revealed many data on the composition of fungus communities, periodicity and constancy of fruit bodies. Some forest communities prove to be very rich in species, for instance up to 800–1200 species in some deciduous virgin forest types in Bialowiecza, Poland (Falinski & Mulenko, 1995). Experimental work has recently shed some light on the population dynamics of fungi, including the distribution, size and age of individual mycelia (e.g. Dahlberg & Stenlid, 1990). Field experiments have shown that an impoverished ectomycorrhizal biota, caused by air pollution, can be partly restored by removal of polluted litter layers (Baar & Kuyper, 1998).

Nevertheless, many fundamental questions remain to be solved. One of the intriguing questions is whether macrofungi, known as rare species, are really rare or only organisms that rarely produce fruit bodies. In conservation practice we accept the former hypothesis, but nobody is sure about it. Theoretically this question can be answered by analysis of mycelium populations with molecular techniques. It has already been demonstrated that the underground population of ectomycorrhizas on roots is not necessarily correlated with the frequency of fruit bodies of various species in the same plot. This may also be true for saprotrophs.

Another question is to what extent the enormous species diversity of fungi also represents functional diversity. For field mycologists many species seem to occur in the same habitat on the same substrate, using the same type of habitat exploitation. For instance: the numerous ectomycorrhizal species in a homogeneous forest plantation, the diversity of Aphyllophorales on branches and twigs, the surprisingly high number of fungi on a homogeneous substrate like dung. And why are some dung-inhabiting fungi exceedingly rare everywhere, whereas others are widespread and common?

A third question concerns the reproduction strategies of fungi. Are the billions of spores really necessary to maintain populations in the long run or is there a great surplus? How effective is fungus colonisation by spores, for instance after long-distance transport by air currents? Unsolved problems like these are relevant for development of a strategy on mushroom harvests and for judging the importance of the ecological infrastructure of a landscape for fungi. The answers to these questions are, in fact, crucial for conservation policies and research in these fields seems to be most rewarding.

The changing mycota

In various publications changes in the mycota are described on a local or national scale, including both evidence of increasing and decreasing species (e.g. Arnolds, 1988; Senn-Irlet, 1997). For conservation purposes data on decrease are the most relevant. Important causes of decline of macrofungi are:

- Habitat destruction, e.g. clear-cut of old-growth forests, digging of peat in peat bogs, building recreational facilities on sand dunes and in alpine areas.

- Changes in land use, often due to EU policy, e.g. replacement of native forest by plantations of exotic trees; improvement, abandonment or afforestation of old, semi-natural meadows.
- Intensification of agriculture, e.g. higher input of fertilisers, deep drainage, both measures with negative effects on adjacent nature areas as well.
- Air pollution, in particular causing acidification and nitrogen enrichment (eutrophication). The decline of ectomycorrhizal fungi in densely populated areas of Europe is mainly caused by these factors (Termorshuizen & Schaffers, 1991), but acidification and eutrophication are without doubt also harmful to some saprotrophs in and outside forests.

Harvesting of edible mushrooms is a much discussed subject in connection with decline (e.g. Leonard & Evans, 1997). In experimental plots in Switzerland (Egli *et al.*, 1990) and the Pacific USA (Pilz & Molina, 1996; and see Chapter 3) no negative effect of intensive picking of fruit bodies has been observed. Nevertheless, representatives of eastern Europe and Portugal expressed great concern on the situation in their countries during ECCF meetings. It is quite possible that destructive harvesting and side effects of collecting, such as trampling, have a negative impact in some regions. Also accurate data on quantities harvested and exported are lacking. It would be a good possibility for European co-operation to initiate a project on the ecological, economic and social aspects of mushroom harvests.

Red Data lists

Red Data lists enumerate species that are considered as threatened or susceptible in a certain area. They are necessarily based on formal or informal data on distribution, ecology and trend of species, so in fact on a compilation of mycological knowledge. In agreement with guidelines of the International Union for Conservation of Nature (IUCN) (Anon., 1995*a*), usually five categories of threatened species are distinguished: extinct, critically endangered, endangered, vulnerable and susceptible or rare.

Since the first Red Data list of fungi appeared in 1982, many European countries have published formal or informal Red Data lists, occasionally also revised editions. Many Red Data lists are difficult to trace since they

Table 4.1. *National Red Data lists of macrofungi in Europe*

Country	Authors	Year	Number of species
Austria	Krisai	1986	211
Austria (2nd edn)	Krisai	1999	542
Czech Republic	Kotlaba *et al.*	1995	120
Denmark	Vesterholt & Knudsen	1990	898
Estonia	Anon.	1995b	76
Finland	Rassi & Väisänen	1987	161
Finland (2nd edn)	Rassi *et al.*	1992	325
Germany East	Benkert	1982	309
Germany West	Winterhoff	1984	1032
Germany (2nd edn)	Benkert *et al.*	1992	1402
Great Britain	Ing	1992	453
Greece	Diamandis	2000	150
Hungary	Rimóczi	1998	535
Latvia	Vimba & Peterans	1996	38
Lithuania	Kutorga *et al.*	1999	740
Macedonia	Karadelev	2000	67
Netherlands	Arnolds	1989	944
Netherlands (2nd edn)	Arnolds & van Ommering	1996	1655
Norway	Bendiksen & Hoiland	1992	649
Norway (2nd edn)	Bendiksen *et al.*	1997	831
Poland	Wojewoda & Ławrynowicz	1986	800
Poland (2nd edn)	Wojewoda & Ławrynowicz	1992	1013
Spain and Portugal	Calonge	1993	153
Sweden	Anon.	1991	514
Sweden (2nd edn)	Aronsson *et al.*	1995	528
Switzerland	Senn-Irlet *et al.*	1997	232
Ukraine	Shelyak-Sosonka	1996	56
USSR (former)	Borodin *et al.*	1984	17
Yugoslavia	Ivancevic	1998	97

were published in rather obscure journals or pamphlets. A survey of national Red Data lists, hopefully up to date, is presented in Table 4.1. A total of approximately 3000 species are included in one or more lists. Regional lists have been omitted, such as the well-documented lists for most German states.

The status of Red Data lists varies from official documents, approved by the government to informal, provisional papers, compiled by a group of mycologists or even a single individual. Also, the applied methods and criteria vary considerably. Most Red Data lists are necessarily based on expert judgement because of lack of reliable data on distribution of species in the past. Only in The Netherlands has an attempt been made to use quantitative criteria (Arnolds & van Ommering, 1996), as prescribed by IUCN (Anon., 1995*a*). In that country a large set of 700 000 computerised data was available. Nevertheless, it was only possible to use the IUCN criteria after adaptation to the specific characteristics of the data set. One third of the species could not be considered because of lack of data. It was also concluded that a Red List based on expert judgement may produce comparable results, in some cases even more realistic results (Arnolds, 1997). I consider the international trend for scientific perfection in Red Data lists as a risk to progress in compilation of Red Data lists of fungi and therefore for conservation of the mycota. It cannot be denied that available databases in almost all European countries are too incomplete to meet the quantitative criteria of IUCN.

The number of species on national Red Data lists varies between 17 in former USSR and 1655 in The Netherlands (Table 4.1). It is evident that these figures are by no means representative of the real threats to the mycota in various countries. Indicative is the difference in the number of listed species in similar countries such as Latvia (38 species) and Lithuania (740 species). The compilation of the lists depends, among other things, on the available data, the applied criteria and the aims of Red Data lists. For instance, in some countries all species on the list are automatically protected by law, but in other countries lists do not have such consequences. In some countries two editions of a Red Data list have already been published. The second edition invariably contains more species than the first. In general, the increase in species is mainly caused by increased knowledge.

In the framework of necessary European co-operation, standardisation of the procedures is badly needed. International criteria should be developed by the ECCF that bridge the gap between rigid IUCN criteria (Anon., 1995a), specific mycological needs and practical possibilities. A preliminary Red Data list for Europe, based on available national lists, was published by Ing (1993). It contains 278 species.

Threatened habitats of fungi

Conservation of fungi is, like conservation of other organisms, in the first place conservation of their habitats combined with adequate management. Conservation of appropriate habitats is also essential for the protection of phanerogams, birds and other organisms that are more popular and better known than fungi. In my opinion we may leave the initiative for the conservation of most habitat types to these specialist groups and national and international organisations with a general conservation goal, such as the IUCN and WWF. If a tropical rainforest is protected for the rhinoceros or tiger, fungi will also automatically profit from the maintenance of the ecosystem. In Europe many efforts are already under way for conservation of habitats such as peat bogs, sea dunes and alpine meadows. The mycological community can support these attempts by delivering additional data on the biodiversity of fungi in those areas and the occurrence of endangered species.

However, mycologists should pay special attention to those habitats that are rich in (threatened) fungal species and that are not adequately characterised by other organisms, in particular plants. In this connection I mention primeval forests, forests on very poor soils and old grasslands. These examples are certainly incomplete, also because of lack of relevant information from large parts of eastern and southern Europe.

Primeval, virgin and old-growth forests share an undisturbed development of vegetation and soil extending over several centuries. In Europe they are found mainly in remote areas in the north and east, elsewhere in a few strict reserves. Characteristic features are abundance of large dead logs in all stages of decay, uneven age of trees, the occurrence of gaps and uneven soil surface caused by holes left by uprooted trees. Characteristic phanerogams are few, but primeval forests are well characterised by a number of lichens, bryophytes and fungi, along with some groups of animals. In Scandinavia and the Baltic states extensive programmes are carried out to locate and preserve valuable old forest relics. Polypores and other lignicolous Aphyllophorales are used as indicator organisms (e.g. Kotiranta & Niemelä, 1993; and see Chapters 5 and 15). Old-growth forests are even rarer and more endangered in the belt of deciduous forests and in the Mediterranean area. Also in these cases characteristic fungi may be used as indicator species, e.g. *Hericium coralloides* for undisturbed beech forests and *Podoscypha multizonata* for very old oak trees.

Both coniferous and frondose forests on very oligotrophic soils, in particular poor in nitrogen, are extremely rich in ectomycorrhizal fungi. Many species are restricted to these forest communities, such as the *Cladonio-Pinetum* (e.g. Wöldecke & Wöldecke, 1990), *Dicrano-Quercetum* (Jansen, 1984), *Luzulo-Fagetum leucebryetosum* and *Carici-Fagetum* (Jahn *et al.*, 1967), the latter type growing on calcareous slopes. Apparently, the slowly growing trees on these sites can only survive with the aid of their numerous mycorrhizal partners. These forest types are very sensitive to eutrophication, caused either by gradual accumulation of nitrogen in forest soils due to air pollution, or by deliberate application of fertilisers in order to increase timber production. Special attention must be given to conservation of these habitats in central and western Europe since nitrogen deposition exceeds critical values in many regions. In strongly polluted areas, such as The Netherlands and northern Germany, oligotrophic forests have become extinct or greatly impoverished and so is their mycota (Arnolds, 1991; Wöldecke, 1998).

A third, strongly threatened habitat type of immense mycological significance is that of old, unfertilised, semi-natural and natural grasslands. These communities are rich in specialised saprotrophs, such as many species of the genera *Hygrocybe*, *Entoloma*, *Clavaria* and *Geoglossum*. They are often indicated as '*Hygrocybe* grasslands'. Old-growth grasslands, parallel to old-growth forests, could be an alternative term. The most characteristic species depend on continuous grassland use over decades or centuries, undisturbed soil and management by grazing or cutting. Some areas are rich in rare phanerogams as well, but many old-growth grasslands do not possess a spectacular flora of green plants. These communities were widespread in traditional cultural landscapes, but are threatened nowadays all over Europe by two opposite trends: intensification of agriculture and abandonment or afforestation. Special attention is paid to the conservation of these grasslands in Scandinavia (e.g. Boertmann, 1995) and Great Britain (see Chapter 10).

All three mycologically important habitat types are very suitable for European co-operation concerning inventories, monitoring and attempts to improve conservation and management. It should be possible to obtain European funding for research on these subjects. But who is taking the initiative?

Fungi in conservation practice in Europe and North America

Important progress has been made in gathering knowledge on the needs of fungal conservation, thanks to only a small number of professional and amateur mycologists who are devoted to the subject. Most people involved are all-round mycologists, usually also involved in research on taxonomy and/or ecology and teaching. Specialists in fungal conservation do not yet exist as they do in other disciplines of biology. Nevertheless, numerous data on the subject are available for anyone interested in publications in scientific and popular journals, proceedings of meetings and Red Data lists. However, the practical impact of this information has been limited up to now. The occurrence of threatened fungi is still rarely used as an important criterion for selection of nature reserves or the application of a certain management. There is little progress in legal protection of fungi, either in national or international legislation. Also, funding of research in favour of fungal conservation lags far behind funding of projects on birds, mammals, phanerogams, butterflies, dragonflies and so on.

The impact of fungi in international nature conservation is still minimal. For instance, the influential International Union for Conservation of Nature (IUCN) has numerous expert groups, covering all living organisms. There are even different expert groups for Indian and African elephants. The entire Kingdom Fungi is represented by only one expert group that has published a few newsletters, but during the last five years it has undertaken no action at all. The organisation Planta Europa was established in the early nineties to co-ordinate and stimulate conservation of plants throughout Europe. Mycologists were not involved and were even unaware of this initiative until some representatives of the ECCF were invited to attend the second congress in Uppsala and to present information on fungus conservation there (Arnolds, 1999). During that congress we learnt much about European legislation on nature conservation, for instance about the importance of the Bern Convention. This convention from 1979, signed by 28 European countries, regulates protection of threatened animals and plants and their habitats. Some mosses and lichens are included, but fungi are completely lacking. At present attempts are being made to have some species protected under these regulations.

Another recent international development is the European Community Directive on the Conservation of Habitats, Fauna and Flora: in short, Habitats Directive. Member states can be obliged to protect areas with great significance for biodiversity. Again, fungi are not considered and mycologists are not involved in this initiative.

In North America mycological knowledge about taxonomy, distribution as well as ecology, is less complete than it is in Europe. Conservation was not on the agenda of mycologists for a long time until harvesting of wild mushrooms became a booming business in the late eighties. In particular in the Pacific Northwest, many studies are currently investigating the influence of harvesting and forest management on the mycota. An anthology of these studies was recently published by Pilz & Molina (1996; and see Chapter 3). The future of old-growth forests is an important political question in this region, which even directly concerned President Clinton. Timely warnings by mycologists have emphasised the crucial importance of fungi in these ecosystems and their suitability as bioindicators. They have successfully influenced management practices and research efforts, as exemplified by this quotation from the report by Pilz & Molina (1996):

> The final record of decision is the document that specifies how the Final Supplemental Environmental Impact Statement (an official governmental document) will be implemented. Table C-3 in that document lists 234 fungus species deemed of 'special concern'. These species were analysed and classified into four survey categories. Survey strategy 1 requires management of known sites for all species shown in Table C-3. Sites with rare and endemic fungus species will have 64.8 hectares (160 acres) temporarily withdrawn from management activities until those sites can be thoroughly surveyed and site-specific management measures prescribed. One particular fungus, the 'noble polypore' (*Oxyporus nobilissimus*), will have 240 hectares (600 acres) of habitat preserved at each known fruiting location. Survey strategy 2 requires surveys before the occurrence of any ground-disturbing activities. Only two fungus species require this strategy: *Oxyporus nobilissimus* and *Bondarzewia montana.* Survey strategy 3 requires conducting extensive surveys to find high-priority sites for species management. All 234 fungus species are included in survey strategy 3, and protocols are currently under development. The general regional surveys of strategy 4 are designed to provide further information about little known species not yet designated as rare and endemic. These regional surveys must begin before 1997 and are expected to take 10 years to complete.

In Europe, mycologists can only dream about efforts on this scale for fungus conservation and research on threatened species!

Conclusions

In the last two decades conservation of fungi has become one of the important duties and fields of research for European mycologists, in particular field mycologists working on macrofungi. At the same time the number of professionals involved in these activities has decreased in most countries. This results in growing responsibilities for fewer people. The practical results of efforts related to fungus conservation are rather limited so far. Financial support for research and public information is also poor. The countries of Fennoscandia are positive exceptions in both respects. I think that the general situation is caused by (1) failure to spread the message sufficiently outside circles of mycologists, (2) inadequate attention to international developments, (3) lack of acquaintance with and unclear status of organisations, such as ECCF, (4) lack of international co-operation in mycological affairs.

I am convinced that fungi deserve, and can gain, more attention in nature conservation, environmental policy and management of nature reserves. Three properties are in favour of this group of organisms: (1) their enormous species diversity and functional importance as decomposers and symbionts, (2) the economic and recreational significance of picking edible mushrooms, and (3) most important of all, the magic of mushrooms: their beauty and variety in shape and colour, their sudden appearance, their role in imagination. Mycologists should make use of the popularity of fungi among people. The present public interest in biodiversity and conservation can and must be used to obtain funding for research and international co-operation, related to conservation of fungi.

The ECCF is a very useful, but informal, expert group without any official status and without funding. It would be better for it to be linked to an authoritative organisation, such as IUCN. Then it would be automatically regarded as an authoritative body and a partner for international treaties. In this connection one might also think of the establishment of a European Mycological Society. Such a society should be the organiser of future Congresses of European Mycologists, another institution without a sound basis in the international scientific arena. It could also play an important role in co-operation and integration of mycological research, e.g. concerning a common checklist, mapping and monitoring. The publication of a European Journal for Mycology might be useful in this respect.

It is obvious that conservation of fungi depends on the input of mycologists. But at present it seems to be also the other way around: the future of mycologists depends on their input in conservation.

References

Aanen, D. & Kuyper, T. W. (1999). Intercompatibility tests in the *Hebeloma crustuliniforme* complex in northwestern Europe. *Mycologia* **91**, 783–795.

Anon. (1991). Floravardskommittén för Svampar. 1991. Kommenterad lista över hotade svampar i Sverige. *Windahlia* **19**, 87–130.

Anon. (1995*a*). *IUCN Red List Categories*. IUCN: Gland, Switzerland.

Anon. (1995*b*). *Red List of Estonian Fungi*. Informal manuscript, Department of Mycology, Institute of Zoology and Botany of the Estonian Academy of Sciences: Tartu.

Anon. (2000). *Verspreidingsatlas. Kaartenbijlage Overzicht van de paddestoelen van Nederland, deel 1, 2.* Nederlandse Mycologische Vereniging: Baarn.

Arnolds, E. (1988). The changing macromycete flora in The Netherlands. *Transactions of the British Mycological Society* **90**, 391–406.

Arnolds, E. (1989). A preliminary Red Data List of macrofungi in The Netherlands. *Persoonia* **14**, 7–125.

Arnolds, E. (1991). Decline of ectomycorrhizal fungi in Europe. *Agriculture,*

Ecosystems and Environment **35**, 209–244.

Arnolds, E. & Kreisel, H. (1993). Conservation of fungi in Europe. In *Proceedings of the Second meeting of the European Council for the Conservation of Fungi at Vilm, 13–18 September 1991*. Ernst-Moritz-Arndt Universität: Greifswald.

Arnolds, E. (1997). A quantitative approach to the Red list of larger fungi in The Netherlands. *Mycologia Helvetica* **9**, 47–60.

Arnolds, E. (1999). Conservation and management of fungi in Europe. In *Planta Europa, Proceedings of the Second European Conference on the Conservation of Wild Plants* (ed. H. Synge & J. Akeroyd), pp. 129–139. Swedish Threatened Unit: Uppsala, and Plantlife: London.

Arnolds, E., Kuyper, T. W. & Noordeloos, M. E. (1995). *Overzicht van de Paddestoelen in Nederland*. Nederlandse Mycologische Vereniging: Wijster.

Arnolds, E. & van Ommering, G. (1996). *Bedreigde en kwetsbare paddestoelen in Nederland. Toelichting op de Rode Lijst*. IKC Natuurbeheer: Wageningen.

Aronsson, M., Hallingbäck, T. & Mattson, J.-E. (1995). *Rödlistade växter i Sverige 1995*. Artdatabanken: Uppsala.

Baar, J. & Kuyper, T. W. (1998). Restoration of aboveground ectomycorrhizal flora in stands of *Pinus sylvestris* (Scots Pine) in The Netherlands by removal of litter and humus. *Restoration Ecology* **6**, 227–237.

Bendiksen, E. & Hoiland, K. (1992). Red list of threatened macrofungi in Norway. In *Directorate for Nature Management, Report 1992–6*, pp. 31–42. Direktoratet for Naturforvaltning & Fungiflora: Oslo.

Bendiksen, E., Hoiland, K., Brandrud, T. E. & Jordal, J.B. (1997). *Truete og sarbare sopparter i Norge – en kommentert Rodliste*. Direktoratet for Naturforvaltning & Fungiflora: Oslo.

Benkert, D. (1982). Vorläufige Liste der verschollenen und gefährdeten Grosspilzarten der DDR. *Boletus* **6**, 21–32.

Benkert, D. Dörfelt, H., Hardtke, H.-J., Hirsch, G., Kreisel, H., Krieglsteiner, G. J., Lüderitz, M., Runge, H., Schmid, H., Schmitt, J. A., Winterhoff, W., Wöldecke, K. & Zehfuss, H.-D. (1992). *Rote Liste der gefährdeten Grosspilze in Deutschland*. Deutsche Gesellschaft für Mykologie e.V., Naturschutzbund Deutschland e.V. IHW-Verlag: Eching.

Boertmann, D. (1995). *The Genus Hygrocybe. Fungi of Northern Europe 1*. Svampetryk: Greve, Denmark.

Borodin, A. M., Bannikov, A. G. & Sokolov, V. E. (eds) (1984). *Red Data Book of the USSR. Rare and Endangered Species of Animals and Plants*, 2nd edn. Lesnaya Promyshlennost: Moscow.

Calonge, F. D. (1993). Hacia la confección de una lista roja de macromycetes (hongos) en la península Ibérica. *Boletín de la Sociedad Micológica de Madrid* **18**, 171–178.

Courtecuisse, R. & Duhem, B. (1994). *Guide des Champignons de France et d'Europe*. Delachaux et Niestlé: Lausanne.

Dahlberg, A. & Stenlid, A. J. (1990). Population structure and dynamics in *Suillus bovinus* as indicated by spatial distribution of fungal clones. *New Phytologist* **115**, 487–493.

Dennis, R. W. G., Orton, P. D. & Hora F. B. (1960). New checklist of British agarics and boleti. *Transactions of the British Mycological Society* **43**, Supplement.

Diamandis, S. (2000). List of threatened macrofungi in Greece. *European Council for the Conservation of Fungi Newsletter* **10**, 12–14.

Egli, S., Ayer, F. & Chatelain, F. (1990). Die Einfluss des Pilzsammelns auf die Pilzflora. *Mycologia Helvetica* **3**, 417–428.
Falinski, J. B. & Mulenko, W. (1995). Cryptogamous plants in the forest communities of Bielowieza National Park. *Phytocoenosis* **7** (N.S.), 1–176.
Hallingbäck, T. (1994). *Ekologisk Katalog över Storsvampar*. Swedish Threatened Species Unit: Uppsala.
Ing, B. (1992). A provisional Red Data list of British fungi. *Mycologist* **6**, 124–128.
Ing, B. (1993). Towards a Red List of endangered European macrofungi. In *Fungi in Europe: Investigation, Recording and Conservation* (ed. D. N. Pegler, L. Boddy, B. Ing & P. M. Kirk), pp. 231–237. Royal Botanic Gardens: Kew.
Ivancevic, B. (1998). A preliminary Red List of the macromycetes of Yugoslavia. In *Conservation of Fungi in Europe, Proceedings of the 4th meeting of the European Council for Conservation of Fungi* (ed. C. Perini), pp. 57–61. Università degli Studi di Siena, Dipartimento Biologia Ambientale: Siena, Italy.
Jahn, H., Nespiak, A. & Tüxen, R. (1967). Pilzsoziologische Untersuchungen in Buchenwälder des Wesergebirges. *Mitteilungen Floristisch-Soziologische Arbeitsgemeinschaft Neue Folge* **11/12**, 159–197.
Jansen, A. E. (1984). Vegetation and macrofungi of acid oakwoods in the north-east of The Netherlands. *Agricultural Research Reports 923*. Pudoc: Wageningen.
Jansen, A. E. & Ławrynowicz, M. (1991). Conservation of Fungi and Other Cryptogams in Europe. Łódź Society of Science and Arts: Łódź, Poland.
Karadelev, M. (2000). A preliminary Red List of macromycetes in the republic of Macedonia. *European Council for the Conservation of Fungi Newsletter* **10**, 7–11.
Kotiranta, H. & Niemelä, T. (1993). Uhanalaiset käävät Suomessa (Threatened polypores in Finland). *Vesija Ympäristähallinnon Julkaisuja-Sarja (Helsinki)* **B17**, 1–116.
Kotlaba, F., *et al.* (1995). Červena kniha ohrazenych a vzácnych druhov rastlin a zivočichov ČR a SR 4. Sinice a rasy, houby, lisejniky, mechorosty [Red data book of threatened and rare plant and animal Species of Czechoslovakia, vol. 4, Cyanobacteria and algae, Fungi, Lichens, Mosses]. Priroda: Bratislava.
Krieglsteiner, G. J. (1991). *Verbreitungsatlas der Grosspilze Deutschlands (West)*, Band 1. *Ständerpilze*. Ulmer Verlag: Stuttgart.
Krieglsteiner, G. J. (1993). *Verbreitungsatlas der Grosspilze Deutschlands (West)*, Band 2. *Schlauchpilze*. Ulmer Verlag: Stuttgart.
Krisai, I. (1986, 1999). Rote Liste gefährdeten Grosspilze Österreichs. In *Rote Listen gefährdeten Pflanzen Österreichs* (ed. H. Niklfield), pp. 178–192. Bundesministeriums Gesundheit-Umweltschutz: Wien.
Kutorga, E., Urbonas, V. & Gricius, A. (1999). *Checklist of Macrofungi in Lithuania with Evaluation of Threat Status*. Vilnius.
Lange, L. (1974). The distribution of macromycetes in Europe. *Dansk Botanisk Arkiv* **30**, 1–105.
Leonard, P. & Evans, S. (1997). A scientific approach to a policy on commercial collecting of wild fungi. *Mycologist* **11**, 89–91.
Nauta, M. & Vellinga, E. C. (1995). *Atlas van Nederlandse Paddestoelen*. Balkema: Rotterdam.
Pegler, D. N., Boddy, L., Ing, B. & Kirk, P. M. (1993). *Fungi of Europe,*

Investigation, Recording & Mapping. The Royal Botanic Gardens: Kew.
Perini, C. (1998). *Conservation of Fungi in Europe*. Proceedings of the 4th meeting of the European Council for the Conservation of Fungi. Università degli Studi di Siena, Dipartimento Biologia Ambientale: Siena.
Pilz, D. & Molina, R. (1996). *Managing Forest Ecosystems To Conserve Fungus Diversity and Sustain Wild Mushroom Harvests*. US Department of Agriculture, Forest Service, Pacific Northwest Research Station: Portland, Oregon.
Rassi, R., Kaipiainen, H., Mannerkoski, I. & Stähls, G. (1992). *Uhanailaisten aläinten ja kasvien seurantatoimikunnan mietintö [Report on the monitoring of threatened animals and plants in Finland]*. Komiteanmietintö 1991: 30. Ympäristöministeriö: Helsinki.
Rassi, R. & Väisänen, R. (1987). *Threatened Animals and Plants in Finland*. Ministry of Environment: Helsinki.
Rimóczi, I. (1998). Endangered macrofungi and a provisional Red List in Hungary. In *Conservation of Fungi in Europe* (ed. C. Perini), pp. 91–111. Università degli Studi di Siena, Dipartimento Biologia Ambientale: Siena.
Senn-Irlet, B. (1997). Reflections on the conservation of fungi in Switzerland. *Mycologia Helvetica* **9**, 3–18.
Senn-Irlet, B., Bieri, C. & Herzig, R. (1997). A provisional Red List of the endangered larger fungi in Switzerland. *Mycologia Helvetica* **9**, 81–110.
Shelyak-Sosonka, Y. R. (1996). *Chervona Kniga Ukrainy – Roslinniy Svit (Red data book of Ukraine – Plant Kingdom)*. Ukrainskaya Enciklopedia: Kiev.
Termorshuizen, A. & Schaffers, A. (1991). The decline of carpophores of ectomycorrhizal fungi in stands of *Pinus sylvestris* L. in The Netherlands: possible causes. *Nova Hedwigia* **53**, 267–289.
Vesterholt, J. & Knudsen, H. (1990). *Truede storsvampe i Denmark – en rodliste*. Foreningen til Svampekundskabens Fremme: Kobenhavn.
Vimba, E. & Peterans, A. (1996). Latvijas Sarkana gramata. 1. sejums. Senes un kerpji (Red Data Book of Latvia, Vol. 1. Fungi and lichens). Riga.
Winterhoff, W. (1984). Vorläufige Rote Liste der Grosspilze (Makromyzeten). In Rote Liste der gefährdeten Tiere und Pflanzen in der Bundesrepublik Deutschland. 4. Auflage (ed. J. Blab *et al.*), pp. 162–184. Kilda: Greven.
Wöldecke, K. (1998). *Die Grosspilze Niedersachsens und Bremens*. Naturschutz und Landschaftspflege in Niedersachsen **39**, 1–538. Niedersächsisches Landesamt für Ökologie: Hildesheim.
Wöldecke, K. & Wöldecke, K. (1990). Zur Schutzwürdigkeit eines Cladonio-Pinetums mit zahlreichen gefährdeten Grosspilzen auf der Langendorfer Geest-Insel (Landkreis Lüchow-Dannenberg). *Beiträge zur Naturkunde Niedersachsens* **43**, 62–83.
Wojewoda, W. & Ławrynowicz, M. (1986). Red list of threatened macrofungi in Poland. In *List of Threatened Plants in Poland* (ed. K. Zarzyckiego & W. Wojewoda, W.), pp. 47–82. Polska Akademia Nauk: Warzawa.
Wojewoda, W. & Ławrynowicz, M. (1992). Red list of threatened macrofungi in Poland. In *List Of Threatened Plants In Poland*, 2nd edn (ed. K. Zarzycki, W. Wojewoda & Z. Heinrich), pp. 27–56. Polish Academy of Sciences: Cracow.
Zervakis, G., Dimou, D. & Balis, C. (1998). A checklist of the Greek macrofungi including hosts and biogeographic distribution. I. Basidiomycotina. *Mycotaxon* **66**, 273–336.

5

Fungi as indicators of primeval and old-growth forests deserving protection

ERAST PARMASTO

Introduction

Conservation of fungi, particularly populations of rare and endangered fungal species, is only possible when their habitats are protected. Consequently, efforts in fungal conservation will be successful only when co-operation with other environmentalists interested in nature conservation is established. Joint, well-motivated proposals to create new protected areas or to regulate a conservation regime in existing reserves will then accepted more easily than otherwise. The role of mycologists is not only to study the mycota of nature reserves, but also to take the initiative in creating new reserves.

Protection of primeval and old-growth forests

Primeval forests are very rare in Europe. Some kind of forest management has been practised almost everywhere except in those forest stands that are low in productivity and poor in species composition growing on oligotrophic *Sphagnum* bogs in northern areas. Seminatural old-growth forests with minimal human impact have survived mainly in nature reserves.

In Estonia, of the 138 hemerophobic vascular plant species found, 90 are growing only or mainly in forests (Kukk, 1999; Trass, Vellak & Ingerpuu, 1999); there are 79 hemerophobic bryophytes and 88 lichens in the Estonian forests. In the boreal zone, old unmanaged forest is the main refugium for rare and endangered species, including fungi. Protection of different forest site types is effective for conservation of species as well as of communities.

In the European Union, the main target of nature protection has been different groups of higher organisms (primarily birds) and a few selected habitat types (wetlands, Mediterranean communities). Primeval or old-growth forests have been included in nature reserves in earlier times; most of the European forests belong to nongovernmental owners and are intensively managed. In several countries of eastern and Central Europe, ownership has changed during the last decade. The forests are being privatised. Due to the economic difficulties in these countries, the scale of forest cutting has increased considerably. At the same time, due to the influence of European ideology, nature protection has received much more attention than earlier. Among other states, in Estonia the NATURA 2000 plan for the creation of a network of protected sites has obtained priority. Following the political decision of the Estonian Parliament, the intention is to increase the area of strictly protected forests to at least 4% (equivalent to 80 000–90 000 ha).

Inventory of forests in Estonia

At the beginning of 1900s about 14% of the area of Estonia was covered with forest stands and this increased to 50.1% (2 187 700 ha) during the century (Ratas & Raukas, 1997; Kohava, 2000). The cause of the increase is a combination of afforestation of land unsuitable for agriculture in the last fifty years and the drastic decline of agriculture that has occurred in the last decade. Half of the forest area (about 1 000 000 ha) belongs to the State; 2.7% of the total area of Estonia is dedicated to four national parks and nature reserves make up about 1.4%. In total, about 11% of Estonia has some kind of nature protection (Randla, 1996). About 3% of forests is in nature reserves, but this area also includes fens, peat bogs and young forests.

A further half of Estonian forests has been or will be returned to private owners. Due to the economic difficulties in the country and lack of interest in sustainable forestry by most of the owners, coupled with the liberal government policy, this second half is subjected to unlimited cuttings even now.

Twenty-two different types of site and 71 types of forest have been distinguished in Estonia including boreal and nemoral forests. Because of intensive forest management, old-growth natural forests can rarely be seen in this country; intact primeval forests cover only 0.2% of the protected forest area (Randla, 1996). In Tartumaa County, 60% of the old-growth forests found in 1993–1994 were clear-cut before 2000. Virgin or primeval forests exist only in a few strictly protected nature reserves.

In the last two to three years, two new co-ordinated projects have been initiated. The intention of an Estonian–Danish project is to create a Forest

Conservation Area Network in state forests, while the other is an Estonian–Swedish project, Inventory of Woodland Key Biotopes (Habitats). One of the aims is to increase the strictly protected forests from 3% to 4% of the total forest area including all the main but also the rare forest types.

Inventory of forests deserving protection

General approach

A preliminary list of what are presumed to be valuable forest stands in state forests has been compiled for the Network Project using the Forestry database at the Estonian Forest Survey Centre, digitised forest maps and aerial photographs. Rare forest types with an age of at least 70 years are growing on 20 455 ha; all state forests pre-selected for further study together cover an area of about 267 300 ha (K. Viilma & J. Öövel, unpublished). These stands were studied in nature by foresters and sites selected by them were then studied by a group of biologists including botanists (as well as bryologists and lichenologists), a mycologist, an entomologist and (when additional data were needed) ornithologists. In 1998–2000, 29176 ha of forests were considered to be deserving of protection. The forest authorities exclude any management in the selected stands until a final decision on their fate has been made.

In nongovernmental forests, about 170 000 ha has been pre-selected to find key biotopes; during the first year of the pilot field inventory, about 670 habitats (covering 1247 ha) deserving protection were found (Andersson, Ek & Martverk, 1999). Trained foresters carried out the inventory. Several field courses have been organised to train these personnel, especially including identification of Estonian Red Data list species and indicator species. This inventory of key habitats will be continued in all forests regardless of ownership over the next few years.

Mycological studies

For inventory of forests deserving protection, and of key biotopes, fungi may be used as good indicators of the species richness of the biota. In contrast to vascular plants, the species richness of fungi, many microorganisms and insects depend heavily on the forest management type, age of forests and amount of coarse woody debris. In boreal forests, the number of wood-rotting species of fungi is up to four times higher on old decomposed fallen trunks compared with freshly felled trunks (Niemelä,

Renvall & Penttilä, 1995). The number of rare fungal species has been demonstrated to be four times less in managed forests than in key biotopes in Norway (Stokland, Larsson & Kauserud, 1997). My own data indicate that in Estonia there are about five times fewer rare fungi in managed forests. At late stages of wood decay, the trunks maintain 'exceptionally diverse species combinations' of saprotrophic fungi (Renvall, 1995). The highest numbers of wood-inhabiting fungal species, including Red Data list species, were found in old forests with numerous dead logs in Central Norway (Høiland & Bendiksen, 1997). More than a hundred rare species of Polyporaceae and Corticiaceae *sensu lato* have been found only in unmanaged old forests in Estonia (E. Parmasto, unpublished).

Diversity of species composition of fungi is a sign of general biodiversity in a site. In many cases, stands in Estonia with a rich fungal biota were found to have several rare species of vascular plants, insects and birds.

To characterise the level of biodiversity of a site, the presence of Red Data list species has been used for the Woodland Key Biotopes (Habitats) Project in Latvia, Sweden and Estonia. In The Netherlands, the mycological value of sites has been evaluated using a weighted sum of Red Data list species occurring in a site (Jalink & Nauta, 1999). However, there is a big difference between the Red Data lists in Estonia and in most other countries. In Estonia, 91 species of fungi have been included in the Red Data list, in Finland 325, Sweden 528, Norway 758, Denmark about 900, Poland 1013, The Netherlands 1655, and former German Federal Republic 1729. In several countries, the aim of including species in a Red Data list has been the continuation of a 'floristic' inventory of species under the modern slogan of 'Biodiversity Studies'.

A limited number of rare or very rare species recognisable by non-specialists has been included in the Red Data list in Estonia. The intention is to give these species some practical protection, which is not mere formal protection status 'on paper', and to inform people that fungi are worthy of protection equally with other species of living beings. When the Estonian Red Data list of fungi is used for evaluation of old-growth forests or key habitats, the number of forest stands worthy of protection will be much smaller than in countries where ten times more species have been included on their equivalent list.

Use of indicator species

Karström in Sweden, and Kotiranta and Niemelä in Finland (references below) initiated the use of fungi as indicators of old-growth natural forests in 1992–1993. This has been mainly based on qualitative observations not supported by quantitative studies. In Sweden, only five indicator species associated with old-growth forests were used by Karström (1992*a*,*b*). Kotiranta & Niemelä (1993, 1996) list 34 species characteristic for old forests and 23 for virgin forests. In Latvia, ten indicator species were listed by Meiere (1996); in former Yugoslavia, 36 indicator species have been selected by Tortic (1998). A list of 43 indicator species for Estonia was published by Erast and Ilmi Parmasto in 1997; the number of suitable species has increased to 50–55 in 1999.

Selection of indicator species is based on different species distribution influenced by different ecological conditions (Bondartseva, 1999), and correlation between presence of indicator species and general high species diversity. Høiland & Bendiksen (1997) have demonstrated that the highest species diversity of fungi is found in forests with access to logs with high levels of decay and large dimensions lacking in managed forests. The old-growth forest indicator fungi are more or less easily recognisable macrofungi associated mainly or only with old-growth natural (primeval) forests minimally affected by forest management and other kinds of human impact. Indicator species indicate species richness in such areas; that is, they are indicators of natural biodiversity (Parmasto & Parmasto, 1997). The list of indicator species is specific for any region, and may be specific for different types of forest site.

In a list of indicator species, only a few, if any, agarics (gilled mushrooms) and boletes have been included. To describe the species composition of agarics and boletes in a forest stand, the place must be visited more than once a year (cf. Kirby, 1988). Tofts & Orton (1998) published a survey on a 20-year study of the mycota of a nature conservation area in Britain (about 3000 ha). During the 20 years, the accumulative curve of species found is a straight ascending line. This is one reason why polypores and other wood-rotting fungi must be preferred in a short-time inventory; species with perennial fruit bodies are the best for this purpose.

The number of indicator species selected for fieldwork ought to be high enough for there to be a reasonable chance they can be found in natural forests that are candidates for protection. Five species, as proposed by Karström (1992*a*,*b*), is obviously too low a number. These species are quite rare and distributed randomly. A greater number may be suitable for experienced mycologists but may not be practical for foresters without special training. For the latter, species with brightly coloured or otherwise morphologically visually striking fruit bodies are preferred. To limit the number of species, very rare species are excluded.

Table 5.1. *List of indicator species for Estonia*

Aphyllophorales, polypores	Aphyllophorales, other groups
Ceriporia purpurea	*Aleurodiscus amorphus*
Ceriporia reticulata	*Amylocorticium subincarnatum*
Diplomitoporus flavescens	*Asterodon ferruginosus*
Fomitopsis rosea	*Clavariadelphus pistillaris*
Ganoderma lucidum	*Clavariadelphus truncatus*
**Hapalopilus salmonicolor*	*Dentipellis fragilis*
Leptoporus mollis	**Gomphus clavatus*
Oligoporus placentus	*Hericium coralloides* (syn.: *H. clathroides*)
Perenniporia medulla-panis	**Hydnellum* sp. (one species*)
Perenniporia subacida	**Lindtneria trachyspora*
Phaeolus schweinitzii	*Multiclavula mucida*
Phellinus chrysoloma	*Phellodon* sp.
Phellinus ferrugineofuscus	*Phlebia centrifuga*
Phellinus ferruginosus	*Phlebia subochracea*
Phellinus nigrolimitatus	*Pseudomerulius aureus*
Phellinus populicola	*Sarcodon* sp. (excluding *S. imbricatum*)
Physisporinus vitreus	*Serpula himantioides*
Pycnoporellus fulgens	*Sistotrema raduloides*
Rigidoporus crocatus	**Steccherinum robustior*
Skeletocutis odora	*Tomentella crinalis*
Skeletocutis stellae	*Xylobolus frustulatus*
Boletales	Gasteromycetes
**Boletus satanas*	*Geastrum* sp. (8 rare, one common species)
Heterobasidiomycetes	Ascomycetes
**Eocronartium muscicola*	**Sarcosoma globosum*
**Tremiscus helvelloides*	**Sowerbyella* sp. (3 species)
	**Xylaria polymorpha*

* Species included in the Estonian Red Data list.

In Estonia, indicator species have been employed since 1995 for evaluation of forests in two nature reserve areas, and for the general inventory of natural forests since 1998. During the years studied by mycologists, 175 of these were recorded as deserving protection. In most cases, three to five indicator species were found in one selected forest stand, the record number being 14.

Now the use of indicator fungal species is included in the list of standard methods in use by the Forest Conservation Area Network Project. In the Woodland Key Biotopes (Habitats) Project, indicator species combined with the Red Data list species are in use.

In the list shown as Table 5.1, 44 species and five genera of fungi are mentioned. The selection is based on 50-year observations in Estonian forests. The five genera included are mainly comprised of more or less rare indicator species that are not easily distinguishable by a nonmycologist.

Acknowledgements

I acknowledge the valuable personal communications of Kaili Viilma, M. Sc., and Jürgen Öövel (Forest Conservation Area Network Project, Tartu). This research was supported in part by the Estonian Science Foundation (grants nos 2145 and 4086).

References

Andersson, L., Ek, T. & Martverk, R. (1999). *Inventory of Woodland Key Habitats*. Final report. National Forestry Board, Estonia: Tallinn & County Forestry Board: Östra Götaland, Sweden.

Bondartseva, M. A. (1999). Wood-destroying basidial macromycetes in ecosystems of the different state: strategies of propagation and survival. In *Abstracts, XIII Congress of European Mycologists*, p. 16, Alcalá de Henares (Madrid).

Høiland, K. & Bendiksen, E. (1997). Biodiversity of wood-inhabiting fungi in a boreal coniferous forest in Sør-Trøndelag County, Central Norway. *Nordic Journal of Botany* **16**, 643–659.

Jalink, L. M. & Nauta, M. M. (1999). Towards a better protection of mycological valuable sites in The Netherlands: how to detect and manage them. In *Abstracts, XIII Congress of European Mycologists*, p. 61, Alcalá de Henares (Madrid).

Karström, M. (1992*a*). Steget före-presentation. *Svensk Botanisk Tidskrift* **86**, 103–114.

Karström, M. (1992*b*). Steget före in det glömda landet. *Svensk Botanisk Tidskrift* **86**, 115–146.

Kirby, K. J. (1988). The conservation of fungi in Britain. *Mycologist* **2**, 5–7.

Kohava, P. (2000). *Estonian Forests 1999*. Eesti Metsakorralduskeskus: Tallinn. [In Estonian.]

Kotiranta, H. & Niemelä, T. (1993). Uhanalaiset käävät Suomessa. *Vesi-ja Ympäristöhallinnon Julkaisuja-Sarja* **B17**, 1–116.

Kotiranta, H. & Niemelä, T. (1996). *Uhanalaiset käävät Suomessa*. Toinen, uudistettu painos (with English summary: *Threatened polypores in Finland*. Second revised edition). Edita: Helsinki.

Kukk, T. (1999). *Vascular Plant Flora of Estonia*. Teaduste Akadeemia Kirjastus: Tartu & Tallinn. [In Estonian, with a summary in English.]

Meiere, D. (1996). Polypores as bioindicators in Mezole. In *Fungi and Lichens in the Baltic Region*. Abstracts, 13th International Conference on Mycology and Lichenology, Riga (ed. E. Vimba), p. 38–39. University of Latvia: Riga.

Niemelä, T., Renvall, P. & Penttilä, R. (1995). Interactions of fungi at late stages of wood decomposition. *Annales Botanici Fennici* **32**, 141–152.

Parmasto, E. (1999). Fungi as indicators of primeval forests deserving protection. In *Abstracts, XIII Congress of European Mycologists*, p. 102, Alcalá de Henares (Madrid).

Parmasto, E. & Parmasto, I. (1997). Lignicolous Aphyllophorales of old and primeval forests in Estonia. 1. The forests of northern Central Estonia with a preliminary list of indicator species. *Folia Cryptogamica Estonica* **31**, 38–45.

Randla, T. (1996). Nature conservation. In *Estonian Environment: Past, Present and Future* (ed. A. Raukas), pp. 86–88. Ministry of the Environment of Estonia: Tallinn.

Ratas, R. & Raukas, A. (1997). *Main Outlines of Sustainable Development in Estonia.* Ministry of the Environment: Tallinn.

Renvall, P. (1995). Community structure and dynamics of wood-rotting Basidiomycetes on decomposing conifer trunks in northern Finland. *Karstenia* **35**, 1–51.

Stokland, J. N., Larsson, K.-H. & Kauserud, H. (1997). The occurrence of rare and red-listed fungi on decaying wood in selected forest stands in Norway. *Windahlia* **22**, 85–93.

Tofts, R. J. & Orton, P. D. (1998). The species accumulation curve for agarics and boleti from a Caledonian pinewood. *Mycologist* **12**, 98–102.

Tortic, M. (1998). An attempt to a list of indicator fungi (Aphyllophorales) for old forests of beech and fir in former Yugoslavia. *Folia Cryptogamica Estonica* **33**, 139–146.

Trass, K., Vellak, K. & Ingerpuu, N. (1999). Floristical and ecological properties for identifying of primeval forests in Estonia. *Annales Botanici Fennici* **36**, 67–80.

6

Recognising and managing mycologically valuable sites in The Netherlands

LEO M. JALINK & MARIJKE M. NAUTA

Introduction

Until recently, nature conservation in The Netherlands has focused on the protection of plants and birds; management plans for nature reserves contain no measures aiming at maintenance or improvement of the mycota. The reasons for this were mainly practical: although a great deal is known about the distribution, ecology and management of higher plants and birds, there is hardly any knowledge concerning fungi. However, site management that is good for birds and higher plants is not necessarily beneficial for other groups of organisms, such as insects or fungi. Moreover, sites with valuable vegetation are not necessarily rich in fungi, and sites with an abundance of very rare fungi can occur among vegetation of little interest.

Since the Dutch Mycological Society started a project for the recording of macrofungi in 1980, much more information about the distribution and ecological preferences of fungi has become available. The Dutch Mycological Society's database now contains more than a million records. However, the data are not of direct use to managers of nature reserves. The data must be filtered and interpreted to be of practical value.

In an attempt to raise interest in mycota in conservation management plans the Dutch Mycological Society organised a meeting entitled 'Fungi and Nature Conservation, Implications for Management' in 1993, to which nature managers were also invited (Kuyper, 1994). The abstracts of this symposium have been published in Dutch and are widely distributed between organisations and individuals involved in nature management. Apart from expressing concern for fungi the impact of these abstracts has been rather limited. The abstracts were not translated into practice and hardly any change in management was observed.

Interviews established that among the reasons for unchanged practices were that wardens, rangers and managers of reserves have many obligations and limited time, the mycological value of their own reserves is little known and no site-specific instructions were given. Evidently, managers need clear instructions, and mycologically valuable sites should be indicated on maps as precisely as possible. As an example of a site in which more was achieved, a map of the nature management plan for the 'Huys ten Donck' estate does have mycologically valuable sites indicated on it.

One of the positive results of the meeting and its abstracts was that some organisations for nature management requested that the mycologically valuable sites in their nature reserves be indicated, and asked for advice concerning management of these sites. In order to comply with these requests, the Committee for Fungi and Nature Conservation of the Dutch Mycological Society has initiated the project 'Mycological Reserves'.

The project 'Mycological Reserves'

The aim of the project is to improve understanding of, and concern for, fungi by site managers, namely rangers and wardens of nature reserves, owners of estates, and municipality and county councils. This is achieved by indicating mycologically valuable sites, not only within but also outside nature reserves, and by giving advice on site management in order to have fungi better protected.

An index has been developed to define the mycological importance of a site. The mycological value is the weighted sum of Red Data list species occurring in a site. The weighting factor is related to the threat category on the Red Data list (Arnolds & Van Ommering, 1996). A species considered to be seriously threatened is assigned four points, threatened three, vulnerable two and susceptible one point. The total of these species-specific threat scores is then a measure of the mycological value of the site.

The first part of the project aims at an overview of the richest sites in The Netherlands. In search of the Dutch 'mycological crown jewels', the project makes use of the database of the Dutch Mycological Society, although this is organised according to map grid-unit, and no site names are included. Another problem is that as new checklists are published (e.g. Arnolds, Dam & Dam-Elings, 1995), they bring about changes in interpretation of threat categories for taxa in the database and give new meaning to others, thus creating some confusion in the database. A solution to these problems is a recently devised digital checklist containing data such as Red Data list status, and occurrence in The Netherlands, with coded ecological preferences for every species found in The Netherlands (Anon., 1997). By combining the distribution database with the digital checklist, a new database was produced with relevant statistics per square kilometre including: number of records, number of species, number of Red Data list species per category. This automatically calculates the mycological value as defined above. Further, figures are given for two periods: from the oldest record in the database (1850) until 1985, and then 1986 to the present.

Subsequently, a map could be created, indicating the square kilometre grid units with high mycological values. The association of the name of a site with data expressed per square kilometre grid unit can only be done with the help of people in the field. There can be several named sites per grid unit, as well as single sites occupying more than one square kilometre. Combining detailed ordnance survey maps with the known ecological preferences of the observed fungi, it should be possible to relocate named sites with a valuable mycota. Fortunately, the members of the Dutch Mycological Society are very co-operative in finding the sites and their appropriate names.

Results

As a result, several mycologically valuable sites have been rediscovered which had not been visited at all in the last 20 years, an amazing fact in such a well-investigated country. In other sites, once known as hot spots for rare fungi, the mycota has declined, and sometimes the causes for this could be indicated. A list of the 200 best sites has been published (Jalink, 1999). This list has already proved to be valuable since it has attracted the interest of the owners and managers of sites included on it and in some cases this has proved to be an incentive for improved management to the benefit of fungi of the sites.

All information is now being verified, by encouraging fellow mycologists to visit the forgotten or poorly investigated sites, and report on their state of management. Analysing the results confirms again that the mycota in general is declining, something which has also been shown before (Arnolds & Jansen, 1992; Nauta & Vellinga, 1993, 1995). Of the top 25 sites, many have a lower mycological value after 1985 when compared with the period before (Table 6.1). The best site in The Netherlands according to the definition used here is 'Nijenrode', an old estate on clay along the river Vecht, which is fortunate to have a manager who is an enthusiastic and knowledgeable mycologist. But many other sites are not managed for

Table 6.1. *The top 25 mycological sites in The Netherlands*

Site name	Mycological value index 1850–1998	Change between pre-1985 and post-1985[a]	Number of Red Data list species	
			1850–1985	1986–1998
Nijenrode	623	=	129	240
Neerijnen	518	-	145	122
Staverden: Leemputten	402	–	134	64
Bunderbos	373	++	47	171
Tiendweg	372	–	124	21
Gunterstein	357	–	115	53
Groeneveld	344	–	120	11
Schaelsberg	342	=	54	131
Notenlaan	335	–	116	33
Amelisweerd	306	–	115	33
Eikenhorst	296	-	80	49
Twickel	268	–	80	31
Stiphout	263	=	12	114
Duin en Kruidberg	248	–	88	38
Valkenswaard	239	–	83	26
Fort Rhijnauwen	238	=	52	67
Zuiderheide	232	-	71	58
Amelisweerd	230	–	71	38
Stokhem	223	-	62	55
Abbertse bos	217	-	80	41
Beatrix kanaal	213	–	80	5
Koningshof	206	–	65	28
Michielsgestel	202	–	70	0
Huys ten Donck	201	–	83	24
Voornes duin	196	=	52	47

[a]key: –, decline >50%; -, decline 10–50%; =, change −10% to +10%; +, increase 10–100%; ++, increase >100%.

the benefit of fungi, and sometimes site management is detrimental to fungi.

Considering the geographical distribution of sites with the highest number of Red Data list species, it is shown that many of these sites are in the province of Utrecht and occur along the coast. The probable reason for this distribution is the presence of calcareous soils combined with old estates and avenues or relatively old woods, which can be found on clayey soil in the province of Utrecht, and in the inner dunes along the coast and in south Limburg.

Table 6.2. *Guidelines for mycological management: protecting fungi on valuable road side verges*

Do	remove leaf litter
	mow and remove the hay
	use mycorrhizal trees when replanting, such as *Quercus*, *Betula*, *Tilia*
	use nutrient-poor loamy or sandy soil if in-filling is necessary
Don't	use fertilisers
	use lawnmowers or garden shredders
	add nutrient-rich (organic) soil
	spread organic debris and mud from ditches on the road side verge
	remove all trees at once (but instead replant gradually and preferably with the same kind of trees)

Table 6.3. *Guidelines for mycological management: protecting fungi in old forests in estates and parks*

Do	remove leaf litter from mossy patches
	leave old trees
	use mycorrhizal species if additional tree planting is necessary
	remove dense undergrowth of species like *Rubus*, *Acer*, *Prunus serotina*
Don't	spread wood chips (from a garden shredder) on the forest floor or on mossy patches
	remove dead trees
	plant too many trees that produce slowly decomposing litter (*Aesculus*, *Quercus*, conifers)
	run heavy vehicles on the forest floor

Management

Concerning the management of mycologically valuable sites, the Committee for Fungi and Nature Conservation is working on practical guidelines of mycological management, grouped by habitat. As examples, the guidelines for two habitats are given here (Tables 6.2 and 6.3), with an overview of management which is thought to be beneficial for fungi (the Dos), and management which is often carried out with good intentions but thought to be harmful for fungi (the Don'ts). These overviews will be published in the form of leaflets and handed out to any person or organisation that is responsible for a valuable site.

The guidelines are not only meant for nature managers, but also for owners of estates, and municipality and provincial councils. Indeed, for everybody who has to deal with management of nature in whatever way. Although most nature managers have good intentions, bad mistakes are sometimes made. For example, using wood chips on a roadside verge with rich Hydnaceous mycota, or dead branches on moss-rich vegetation to prevent people from walking on it. Both measures result in eutrophication, followed by a rapid increase of grasses and nitrophilous plants and declines of both moss flora and mycota.

Discussion

Of course, this method of estimating the mycological value of sites has its limitations. Such an exercise can only be carried out if a carefully evaluated Red Data list already exists, and if there are enough distribution data. Also, the method largely ignores the biodiversity of organisms smaller than about 1 mm, but not enough is known at present to allow a careful evaluation of the mycological value of sites on the basis of their microfungi. Consequently, such consideration is confined to the macrofungi. But by locating and giving guidelines for better management of mycologically valuable sites it is at least a start towards better protection of them.

References

Anon. (1997). *BioBase 1997, Register Biodiversiteit*. Centraal Bureau voor de Statistiek: Voorburg/Heerlen.

Arnolds, E., Dam, N. & Dam-Elings, M. (1995). Standaardlijst van Nederlandse macrofungi 1995. In *Overzicht van de paddestoelen in Nederland* (ed. E. Arnolds, T. W. Kuyper & M. E. Noordeloos), pp. 754–828. Nederlandse Mycologische Vereniging: Wijster.

Arnolds, E. & Jansen, E. (1992). New evidence for changes in the macromycete flora of The Netherlands. *Nova Hedwigia* **55**, 325–351.

Arnolds, E. J. M. & van Ommering, G.(1996). *Bedreigde en kwetsbare paddestoelen in Nederland. Toelichting op de Rode lijst.* IKC Natuurbeheer: Wageningen.

Jalink, L. M. (1999). Op zoek naar de mycologische kroonjuwelen van Nederland 1: De 200 meest waardevolle kilometerhokken. *Coolia* **42**, 143–162.

Kuyper, T. (1994). Paddestoelen en natuurbeheer: wat kan de beheerder? *Wetenschappelijke Mededelingen van de Koninklijke Nederlandse Natuurhistorische Vereniging* 212.

Nauta, M. M. & Vellinga, E. C. (1993). Distribution and decline of macrofungi in The Netherlands. In *Fungi in Europe, Investigation, Recording and Conservation* (ed. D. N. Pegler, L. Boddy, B. Ing & P. M. Kirk), pp. 21–46. Royal Botanic Gardens: Kew.

Nauta, M. M. & Vellinga, E. C. (1995). *Atlas van Nederlandse paddestoelen*. A. A. Balkema Uitgevers B. V.: Rotterdam.

7
Threats to hypogeous fungi

MARIA ŁAWRYNOWICZ

Introduction

The European Council for the Conservation of Fungi, during its fourteen years of activity, focused attention on threats and conservation problems of fungi. As a result of meetings, workshops and discussions, several regional and national Red Data lists as well as a preliminary European Red Data list (Ing, 1993), have been published.

A decline in the number of species of macrofungi has been reported from different European countries (Arnolds & Jansen 1991). Hypogeous species must also be included among the threatened fungi although hypogeous fungi form a special ecological group of macromycetes because their fruit bodies originate and remain underground until they break down naturally.

Elaphomyces is the most common genus in the northern part of Europe and includes several species that are under threat. *Tuber* is the best-known genus of hypogeous fungi in the Mediterranean area. Maturity of the fruit bodies of many species is indicated by a strong pleasant smell. Indeed, the aroma of truffles has attracted attention since ancient times and there are some highly valued species of the genus *Tuber*, such as *T. melanosporum*, *T. brumale* and *T. aestivum*, which have been exploited for consumption since prehistoric times. In the history of western civilisation, truffles were a frequent subject of research by natural philosophers. Because of the high culinary value assigned to these fungi even early scientists like Theophrastus, Dioscorides and Pliny tried to investigate their nature and biology (Pegler, Spooner & Young, 1993).

Numerous hypogeous species are involved in ectomycorrhizal associations with trees. They also have particular requirements concerning type of soil and climatic conditions. They grow in a large variety of habitats from virgin forests to sites in city centres (Hawker, 1954; Ławrynowicz, 1988). As they occur under the ground, mycologists often fail to record them; hypogeous fungi require different methods of searching and special techniques to enable them to be discovered.

The existing data concerning the occurrence and distribution of hypogeous fungi is still limited to scarce collections in various regions of Europe (Ławrynowicz, 1989, 1990). As rare species, they are placed in the regional and national Red Data lists of endangered fungi.

Red Data lists

Threatened species of macromycetes placed in the national Red Data lists have been summarised by Arnolds & de Vries (1993) and these also include hypogeous species.

Red Data lists of hypogeous fungi have been compiled on the basis of data collected from Germany (Benkert *et al.*, 1996), Poland (Wojewoda & Ławrynowicz, 1992), The Netherlands (Arnolds & van Ommering, 1996), Denmark (Vesterholt & Knudsen, 1990), Finland (Rassi & Väisänen, 1987), Sweden (Larsson, 1997), Norway (Bendiksen & Høiland, 1992), the Czech Republic and Slovakia (Kotlaba, 1995) and HELCOM (Helsinki Commission) – a draft working paper dated 27 November 1995 entitled *Red Lists of Macrofungi in the Baltic and Nordic Regions*.

These Red Data lists include hypogeous fungi comprising 14 genera of ascomycetes and 17 basidiomycete genera (Table 7.1). Altogether, 118 species are recorded, 56 ascomycetes (Table 7.2) and 62 basidiomycetes (Table 7.3) are on the lists in one or other threat category. This means that about 50% of all existing hypogeous fungi are threatened in these parts of Europe.

The number of species in each country varies from 93 in Germany to 6 in the Czech Republic and Slovakia taken together (Tables 7.2, 7.3 and 7.4). In fact, the numbers of species included in these lists indicate the quality and quantity of data available and state of our knowledge, rather than the true extent of the real threat to these fungi. The example from Poland illustrates this point: the number of ascomycetes is higher (Table 7.4) because the taxonomic and ecological revision of this group has already been made (Ławrynowicz, 1988), but basidiomycetes are still awaiting study to this depth.

The final columns of Tables 7.2 and 7.3 show the number of countries that list the threatened species. The majority of species are recorded on the Red list of only a single country of the six that were examined (24 of the 56 ascomycetes and 31 of the 62 basidiomycetes). There are no species indicated on all national lists, although two ascomycete species and one

Table 7.1. *Numbers of threatened species in different genera of hypogeous fungi*

Ascomycetes	Number of species	Basidiomycetes	Number of species
Balsamia	3	*Alpova*	3
Barssia	1	*Arcangeliella*	2
Choiromyces	1	*Chamonixia*	1
Elaphomyces	11	*Elasmomyces*	2
Fischerula	1	*Gastrosporium*	1
Genea	6	*Gautieria*	3
Geopora	2	*Hydnangium*	4
Hydnobolites	2	*Hymenogaster*	16
Hydnotrya	5	*Hysterangium*	8
Pachyphloeus	3	*Leucogaster*	2
Picoa	1	*Martellia*	1
Sphaerosoma	1	*Melanogaster*	6
Stephensia	1	*Octavianina*	1
Tuber	18	*Rhizopogon*	9
		Sclerogaster	1
		Stephanospora	1
		Wakefieldia	1
Total	56	Total	62

species of basidiomycete are found on six of the lists (Table 7.5). It is worth noting that these three species: *Pachyphloeus citrinus*, *Hydnobolites cerebriformis* and *Chamonixia caespitosa* all require natural forest conditions. The number of extinct species is fairly high (25). These species (Table 7.6) need to be checked and analysed first.

As we can see from this analysis of Red Data lists, the strategy or inclusion of hypogeous fungi varies between two extremes. The first is that some authors place almost all known hypogeous fungi appearing in the country among the threatened fungi on the basis of their rarity. The second case is that no hypogeous fungi are included on the Red Data list. These extreme viewpoints emphasise the deficiency of our current knowledge and lack of data. The following reasons account for this: (1) the real status of occurrence has not been recognised because only a few, if any, people in the country specialise in this group of fungi and a random collection cannot be established as the basis for the evaluation; (2) it is not possible to verify all the existing data to confirm reliability; (3) taxonomic difficulties and nomenclatural problems in this group require the support of experts.

Table 7.2. *Threat categories of hypogeous ascomycetes in national Red Data lists*

	Country								
Species	D	PL	NL	DK	FIN	S	N	CZ-SK	NCs*
Hydnobolites cerebriformis	3	3	4	3	4	3			6
Pachyphloeus citrinus	0	1	2	4		3	4		6
Balsamia platyspora	1	3	0	3			4		5
Elaphomyces anthracinus	3	1			4	4	3		5
Hydnotrya michaelis	2	3	3	3		2			5
Tuber aestivum	1	0		3		2		4	5
Tuber maculatum	2	3	2		4		2		5
Elaphomyces leveillei		2			4	4	3		4
Genea verrucosa	1	3		3		4			4
Genea hispidula	0		1	3			4		4
Pachyphloeus melanoxanthus	1	1				2	2		4
Tuber rapaeodorum	3	3	0	3					4
Choiromyces meandriformis	2	3		3					3
Elaphomyces aculeatus	0			3		1			3
Genea klotzschii	0	3		3					3
Pachyphloeus conglomeratus	0		4	3					3
Stephensia bombycina	2		3			1			3
Tuber borchii	3		0	3					3
Tuber foetidum	1		0				2		3
Tuber puberulum	2	4	1						3
Tuber rufum		4	4				3		3
Balsamia vulgaris	1	3							2
Elaphomyces papillatus	0	1							2
Elaphomyces reticulatus	3				4				2
Geopora sumneriana	3		4						2
Hydrobotrya cerebriformis	2		4						2
Hydnotrya tulasnei	3		1						2
Sphaerosoma fuscescens	0	4							2
Tuber dryophilum	1	3							2
Tuber excavatum	3		0						2
Tuber mesentericum	0	1							2
Tuber scruposum	1			3					2
Balsamia fragiformis	2								1
Barssia oregonensis		2							1
Elaphomyces asperulus	1								1
Elaphomyces cyanosporus		1							1
Elaphomyces granulatus			1						1
Elaphomyces muricatus			2						1
Elaphomyces mutabilis	1								1
Elaphomyces striatosporus						4			1
Fischerula macrospora						2			1
Genea fragrans			4						1
Genea lespiaultii		3							1
Genea sphaerica	0								1
Geopora cooperi	3								1
Hydnobolites fallax	2								1

Table 7.2. (*cont.*)

	Country								
Species	D	PL	NL	DK	FIN	S	N	CZ-SK	NCs*
Hydnotrya ploettneriana	1								1
Hydnotrya suevica		3							1
Picoa carthusiana	0								1
Tuber brumale	1								1
Tuber exiguum	0								1
Tuber foetidum ss. Hawker	1								1
Tuber macrosporum	0								1
Tuber michailowskiianum			1						1
Tuber murinum	1								1
Tuber nitidum		4							1
Total 56	43	25	21	14	5	12	9	1	

* NCs, Number of countries with the species on the national Red Data list.

Ecological remarks

Observations of hypogeous fungi from the Red Data list in Poland lead to the conclusion that it is not enough to protect the locality of hypogeous fungi. Rather, it is necessary to provide them with special conditions, taking into account the requirements of hypogeous fungi, which are often different from that of epigeous macromycetes. Two particular examples are worthy of note.

Tuber mesentericum. The most northeasterly locality for this species in Europe is protected in a nature reserve near Częstochowa (Ławrynowicz, 1988). The locality is situated on top of a hill in a beech–oak–hornbeam forest on calcareous soil. Ascocarps of *T. mesentericum* were collected on a walking trail used every year by thousands of tourists and pilgrims passing by. The vegetation on the trail is reduced to 10% of its normal cover, and after rains drainage water can easily run down the trail taking debris with it and sometimes uncovering the upper part of fruit bodies. This place has been monitored twice a year, but the project is kept secret in order to avoid its destruction.

Barssia oregonensis. The only locality in Europe where this species is known to occur is protected in the Tatra National Park (Ławrynowicz & Skirgiełło, 1984). The locality is situated in a mossy *Picea abies* forest, cut through by a tourist

Table 7.3. *Threat categories of hypogeous basidiomycetes in national Red Data list*

Species	Country								
	D	PL	NL	DK	FIN	S	N	CZ-SK	NCs*
Chamonixia caespitosa	3	0			2	2	2	4	6
Melanogaster tuberiformis	3		3	2		2		4	5
Gautieria morchelliformis	2	3	0	3					4
Alpova diplophloeus	3					2	3		3
Gastrosporium simplex	3	1				1			3
Hydnangium carneum	2	1	0						3
Hymenogaster vulgaris			1				4	4	3
Hymenogaster niveus	3				1		2		3
Hymenogaster tener	3		4				3		3
Hymenogaster arenarius	3		4				3		3
Hymenogaster olivaceus		1	1				4		3
Melanogaster variegatus	3	3	3						3
Octavianina asterosperma	3		2			2			3
Rhizopogon vulgaris	3		0	3					3
Arcangeliella stephensii	3							4	2
Elasmomyces mattirolianus	2				2				2
Elasmomyces krjukowensis						2	2		2
Gautieria otthii	2			3					2
Hymenogaster populetorum	3		3						2
Hymenogaster decorus	3						4		2
Hymenogaster muticus	2						3		2
Hymenogaster luteus			4			2			2
Hymenogaster rehsteineri	3		4						2
Hymenogaster megasporus	3		4						2
Hymenogaster hessei	3		0						2
Hymenogaster verrucosus	3		4						2
Hysterangium stoloniferum	3			3					2
Hysterangium calcareum	3							4	2
Hysterangium separabile	3	2							2
Melanogaster ambiguus	3				4				2
Rhizopogon hawkeri	3		4						2
Alpova klikae	3								1
Alpova rubescens	0								1
Arcangeliella asterosperma				2					1
Gautieria mexicana	2								1
Hydnangium pila	2								1
Hydnangium neuhoffii		0							1
Hydnangium cereum	2								1
Hymenogaster bulliardii	3								1
Hymenogaster cinereus	2								1
Hymenogaster thwaitesii	1								1
Hysterangium thwaitesii	2								1
Hysterangium clathroides					4				1
Hysterangium crassum			1						1
Hysterangium nephriticum	3								1
Hysterangium rickenii	1								1
Leucogaster badius	3								1
Leucogaster nudus	3								1

Table 7.3. (*cont.*)

Species	Country D	PL	NL	DK	FIN	S	N	CZ-SK	NCs*
Martellia soehneri	1								1
Melanogaster broomeianus			3						1
Melanogaster intermedius	3								1
Melanogaster macrosporus	2								1
Rhizopogon briardii	3								1
Rhizopogon luteolus			2						1
Rhizopogon luteorubens			2						1
Rhizopogon marchii	2								1
Rhizopogon parksii	3								1
Rhizopogon rubescens				3					1
Rhizopogon villossus	3								1
Sclerogaster compactus	2								1
Stephanospora caroticolor	3								1
Wakefieldia macrospora	1								1
Total 62	50	8	21	7	5	7	10	5	

*NCs, Number of countries with the species on the national Red Data list.

Table 7.4. *Numbers of hypogeous ascomycetes and basidiomycetes on Red Data lists of different countries*

	Total	D	PL	NL	DK	FIN	S	N	CZ-SK
Ascomycetes	56	43	25	21	14	5	12	9	1
Basidiomycetes	62	50	8	21	7	5	7	10	5

Table 7.5. *Distribution of hypogeous fungi between the national Red Data lists of different countries*

	Total	Number of lists recording the same species 1	2	3	4	5	6
Ascomycetes	56	24	11	9	5	5	2
Basidiomycetes	62	31	17	11	1	1	1

Table 7.6. *Extinct species of hypogeous fungi*

Species	Country	Species	Country
Ascomycetes		Basidiomycetes	
Balsamia platyspora	NL	*Alpova rubescens*	D
Elaphomyces aculeatus	D	*Chamonixia caespitosa*	PL
Elaphomyces papillatus	D	*Gautieria morchelliformis*	NL
Genea hispidula	D	*Hydnangium carneum*	NL
Genea klotzschii	D	*Hydnangium neuhoffii*	PL
Genea sphaerica	D	*Hymenogaster hessei*	NL
Pachyphloeus citrinus	D	*Rhizogogon vulgaris*	NL
Pachyphloeus conglomeratus	D		
Picoa cartusiana	D		
Sphaerosoma fuscescens	D		
Tuber aestivum	PL		
T. borchii	NL		
T. excavatum	NL		
T. exiguum	D		
T. foetidum	NL		
T. macrosporum	D		
T. mesentericum	D		
T. rapaeodorum	NL		

route and with numerous walking trails running through the forest. Ascocarps of *Barssia* grow on both sides of the trail, partly in the upper layer of soil among mosses and liverworts in areas trodden by tourists wandering off the trail to avoid wet places.

As these examples imply, disturbance may encourage hypogeous fruit body formation. Some hypogeous fungi are permanent components of urban ecosystems, inhabiting ruderal places, growing under trees in parks and streets, close to buildings (Ławrynowicz, 1988). Lilian Hawker (1954) wrote: 'Even those forms, such as *Tuber puberulum*, that grow within the layer of partially decayed leaves, are not found where the leaves drift to form a thick layer. Other species, such as *Tuber excavatum*, are found only in places such as edges of woods, artificial mounds or banks or near the top of slopes, where leaves do not accumulate. Slight disturbance of soil may stimulate some species, presumably through improved aeration. Thus, large numbers of young fruit bodies of *T. puberulum* were found on returning to a patch which when searched, and therefore disturbed, some months earlier had yielded only few. *Tuber aestivum* was found in quantity in the grounds of the University of Bristol one year after the ground had been dug, after having been undisturbed for many years.'

For more than 300 years efforts have been made to cultivate truffles extensively in the south of Europe. The most famous truffières are in France. Here we can see the exact place where truffles produce their ascocarps since the vegetation becomes extremely reduced. *Tuber melanosporum*, which yields fruit bodies over many years and sometimes in great quantity, provides an optimistic example for the future of hypogeous fungi.

On the basis of analysis of Red Data lists, together with field observations of several threatened hypogeous fungi, we arrived at the following conclusions about steps necessary for the conservation of hypogeous fungi.

Conservation of habitat. It is necessary to protect the locality of a threatened fungus on a large scale, including potential habitats of threatened species.

Protection of subterranean features of the habitat. The features in need of attention are no accumulation of litter, scanty vegetation cover, and exposure to direct sunlight and occasional running water.

Monitoring of sites. It is necessary to collect the data concerning frequency and abundance of threatened fungi to know the trends and scale of their changes. Habitat requirements of threatened species should be carefully observed also.

Habitat management. By afforestation or by changing the soil structure it is possible to create new sites for hypogeous fungi.

Silent protection. Information on the occurrence of truffles should not be distributed among the local inhabitants because the locality is likely to be too intensively harvested.

References

Arnolds, E. & Jansen A. E. (1991). Conclusions of the First Meeting of the European Committee on the Protection of Fungi, Łódź, August 10–13, 1988. In *Conservation of Fungi and Other Cryptogams in Europe* (ed. A. F. Jansen & M. Ławrynowicz), *Science Tracks* **18**, pp. 90–103. Łódź Society of Science and Arts: Łódź, Poland.

Arnolds, E. & de Vries, B. (1993). Conservation of Fungi in Europe. In *Fungi of Europe: Investigation, Recording and Conservation* (ed. D. N. Pegler, L. Boddy, B. Ing & P. M. Kirk), pp. 211–230. Royal Botanic Gardens: Kew.

Arnolds, E. & Van Ommering, G. (1996). *Bedreigde en kwetsbare paddestoelen in*

Nederland. IKC Natuurbeheer: Wageningen.

Bendiksen, E. & Høiland, K. (1992). *Red list of threatened macrofungi in Norway.* Report for 1992–6, pp. 31–42. Directorate for Nature Management: Oslo.

Benkert, D., Dörfelt, H., Hardtke, H. J., Hirsch, G., Kreisel, H., Krieglsteiner, G. J., Lüderitz, M., Runge, A., Schmid, H., Schmitt, J. A., Winterhoff, W., Wöldecke, K. & Zehfuss, H. D. (unter Mitarbeit von Einhellinger, A., Gross, G., Grosse-Brauckmann, H., Nuss, I. & Wölfel, G. K.) (1996). Rote liste der Grosspilze Deutschlands. *Vegetationskunde* **28**, 377–426. Bundesamt für Naturschutz: Bonn-Bad Godesberg.

Hawker, L. E. (1954). British hypogeous fungi. *Philosophical Transactions of the Royal Society of London, Series B* **237**, 429–546.

Ing, B. (1993). Towards a Red List of Endangered European Macrofungi. In *Fungi of Europe: Investigation, Recording and Conservation* (ed. D. N. Pegler, L. Boddy, B. Ing & P. M. Kirk), pp. 231–237. Royal Botanic Gardens: Kew.

Kotlaba, F. (1995). *Červena kniha ohrazenych a vzácnych druhov rastlin a zivočichov ČR a SR 4. Huby (Makromycéty)*, pp. 30–119. Priroda: Bratislava.

Larsson, K. H. (1997). *Rödlistade svampar i Sverige – Arfakta* [*Swedish Red Data Book of Fungi*]. Art. Databanken, Sveriges Lantbruksuniversitet: Uppsala.

Ławrynowicz, M. (1988). *Flora Polska. Grzyby (Mycota) 18: Ascomycetes; Elaphomycetales, Tuberales.* Państwowe Wydawnictwo Naukowe: Warszawa-Kraków.

Ławrynowicz, M. (1989). Chorology of the European hypogeous Ascomycetes, I. Elaphomycetales. *Acta Mycologia* **25**, 3–41.

Ławrynowicz, M. (1990). Chorology of the European hypogeous Ascomycetes. II. Tuberales. *Acta Mycologia* **26**, 7–75.

Ławrynowicz, M. & Skirgiełło, A. (1984). *Barssia*, a new genus in Europe. *Acta Mycologia* **20**, 277–279.

Pegler, D. N., Spooner, B. M. & Young, T. W. K. (1993). *British Truffles. A Revision Of British Hypogeous Fungi.* Royal Botanic Gardens: Kew.

Rassi, P. & Väisänen, R. (1987). *Threatened Animals and Plants in Finland.* Ministry of Environment: Helsinki.

Vesterholt, J. & Knudsen, H. (1990). *Truede Storsvampe I Denmarken Rødliste.* Foreningen til Svampekundskabens Fremme: København.

Wojewoda, W. & Ławrynowicz, M. (1992). Red list of threatened macrofungi in Poland. In *List of Threatened Plants in Poland*, 2nd edn (ed. K. Zarzycki, W. Wojewoda & Z. Heinrich), pp. 27–56. Polish Academy of Sciences: Cracow.

8
Wild mushrooms and rural economies

DAVID ARORA

Introduction

Wild mushrooms have been praised by Roman and Chinese emperors alike while providing an important everyday food source for rural people around the world. But in America and other English-speaking countries, a deep and exaggerated distrust of mushrooms has denied them a cherished place at the dinner table. As Stephen Jay Gould has pointed out (Gould, 1997), this prejudice is expressed even in our everyday language: urban crime and taxes are said to 'mushroom' while prosperity and the arts 'flower'!

Until recently very few Americans dared to eat wild mushrooms, and even fewer picked them. But in the 1980s, rising demand in the wealthier mushroom-loving countries caused them to look abroad for new sources of wild mushrooms. At the same time, Americans' palates grew bolder and more sophisticated. Entrepreneurs rushed in to fill the rapidly growing demand for gourmet foods, and out-of-work rural Americans and recent immigrants (particularly from southeast Asia) saw picking wild mushrooms as a chance to make a decent living in a familiar environment, the forest, while maintaining a measure of personal dignity and cultural autonomy.

Thousands of people in the Pacific Northwest now pick and sell wild mushrooms. Most of them pick locally or opportunistically for a little extra cash or as one of several seasonally based strategies for survival. But the notoriously fickle nature of mushrooms – they may be overwhelming abundant one year and then frustratingly scarce the next – has created the need for a skilled cadre of pickers and buyers (many do both) willing to go where the rainbows lead them. And the rare combination of abundant public lands and private vehicles makes it possible for whoever so chooses to do just that.

The global mushroom trade

Wild mushrooms have long been gathered intensively in many parts of the world. But with the globalisation of trade, mushrooms are now being picked in more places than ever before and they are travelling farther and faster. The global trade in matsutake alone is estimated at US$3 to 5 billion annually; for chanterelles it is about US$1.5 billion.

In many developing countries wild mushrooms have become an important source of income for people in remote forested regions where there are few other opportunities to make money. Impoverished farmers in Bulgaria, for instance, have bought new tractors with money gained from selling boletes to Italy, and villagers in Zimbabwe pay school tuition fees for their children by selling mushrooms from their native miombo woodlands, including chanterelles that they can ship out of season to Europe.

One particularly dramatic success story is in Champa, on the eastern flank of the Himalayas where the Chinese provinces of Yunnan, Sichuan, and Tibet come together. Virtually all of the Tibetan villagers in this rugged forested region spend the summer months picking, buying, and selling wild mushrooms, or servicing those who do.

Despite a short growing season, the mushrooms provide families with anywhere from 50% to 100% of their annual income. In two months, some Tibetan families living in matsutake-rich forests are able to earn more than ten times the annual average wage of a worker in developed regions of China like Shanghai. And in contrast to North America, where the pickers are widely scattered and the material benefits of the mushroom harvest are difficult to distinguish, the wealth generated by mushrooms in Champa is dramatically evident to outsiders because virtually all of the local money comes from the mushroom trade. Villages near matsutake beds are dotted with new two-storey wooden houses built in the traditional Champan style but several times larger and more ornate than anything known before. Small shops and other businesses have begun to blossom as the new homeowners look for other ways to invest their money.

Even in forests where less valuable mushrooms predominate (e.g. chanterelles or gypsy mushrooms, a locally sold species) the villagers have new houses. They are less grandiose than the matsutake 'mansions', but are still a dramatic improvement over the hovels that characterise deforested areas.

Rural boomtowns financed by distant urban elites are certainly nothing new. But what is extraordinary about this area in Champa is that intact, healthy forests are perceived as the key to rural development rather than as an impediment to it. Several villages have developed their own mushroom management plans, timber harvest has been scaled back, and the indigenous culture is far more intact than in nearby areas that cater to tourism.

A wild mushroom primer

Japan and western Europe are the major importers of wild mushrooms, but many food stores in America now carry a small selection of fresh and dried fungi, and it has become *de rigueur* for up-market urban restaurants to feature morels, chanterelles, boletes (porcini), lobster mushrooms, candy caps, and other kinds, depending on their availability. Here is a rundown on some of the kinds you may encounter.

Matsutake are the most expensive wild mushrooms after truffles. The Asian species, *Tricholoma matsutake*, sells for as much as US$200 a piece in Tokyo markets. The white North American version, *T. magnivelare*, has the same complex, spicy fragrance but is somewhat cheaper. Found across North America and south as far as Guatemala, it is most abundant in the coniferous forests of the Pacific Northwest and the tanoak woodlands of northern California. Nearly all of the crop goes to Japan, but in good years you can find it fresh in America in Japanese restaurants and markets, and in a few gourmet food stores. The Japanese pay phenomenal prices for young specimens, but opened ones are considerably cheaper yet taste just as good.

Chanterelles (*Cantharellus cibarius* and close relatives) are the best known wild mushrooms on the west coast. The lovely, yellow-orange mushrooms have a fruity fragrance and chewy texture, and can be refrigerated for up to a month after picking. They are found throughout the world and are nearly always sold fresh. Up-market restaurants in America serve them practically year-round. The autumn crop from the coniferous forests of the Pacific Northwest is legendary. A slightly different species is abundant in the oak woodlands of California during the winter. In the summer, fresh chanterelles originate from Nova Scotia, Saskatchewan and eastern Europe.

Morels (*Morchella* species) are prized in America as well as western Europe. Their hollow, honeycombed caps are distinctive but blend in uncannily with their surroundings. Morel hunting is a popular pastime in the Midwest, but the really big crops occur after forest fires in the West. The several species that are harvested differ somewhat in flavour and colour. Fresh morels are most plentiful during the spring and summer, but the Mexican crop comes in during the autumn and even into December. Turkey and North America are the major exporters of fresh morels to

Europe, while the Himalayan countries have been exporting large quantities of dried morels since World War II.

The king bolete (*Boletus edulis* and close relatives) is the most popular wild mushroom of Europe. It goes under many names there: porcini, cepe, steinpilz, borovik, etc. West coast king boletes are among the largest in the world, attaining weights of 10 pounds or more, but are very perishable; only limited quantities of young specimens are shipped fresh. Some are frozen whole but most are sliced and dried. Fresh king boletes are served in some up-market restaurants; they usually originate from the coastal and mountain areas of the West, and vary considerably in quality. Dried boletes are famous for their concentrated flavour and wonderful aroma. They are available year-round and may come from almost anywhere, China, Europe, South America, South Africa, etc., but beware of less aromatic imitations.

Hedgehog mushrooms (*Hydnum umbilicatum* and *H. repandum*) are similar to chanterelles in colour and texture, but with a layer of soft spines on the underside of the cap instead of ridges. Mainly used by restaurants during the spring when chanterelles are scarce.

Black trumpets or black chanterelles (*Craterellus cornucopioides*) tend to be difficult to see but look like clusters of black petunias erupting from the ground. The main harvest area is in northern California, but market share is being lost to North Africa and Eastern Europe, which are much closer to the main place of consumption, which is France. This flavourful mushroom has yet to catch on in America, perhaps because of its dark colour. Slow, prolonged cooking is best; excellent fresh or dry.

Lobster mushrooms are becoming increasingly popular in restaurants. This is actually two fungi in one: a fluorescent red-orange fungal parasite (*Hypomyces lactifluorum*) that completely engulfs a brittle white species of *Russula*. Available fresh in the summer and autumn from the Pacific Northwest and Mexico.

Yellow foot (*Cantharellus tubaeformis* and relatives) is a mild-flavoured chanterelle favoured by American chefs for its clean, slim, petite appearance. It is picked on the west coast in winter and spring.

Cauliflower mushroom (*Sparassis crispa*) is a large, fragrant mushroom looking more like a brain than a cauliflower. It grows at the bases of conifers and is best in soups and casseroles.

Candy caps (*Lactarius fragilis*) has become so popular in restaurants that demand exceeds supply. Chefs are enthralled with its powerful aroma. A single cooked or dried candy cap will envelop the entire restaurant in an aroma like maple syrup.

Hen of the woods (*Grifola frondosa*) forms large clusters of small, overlapping caps which look like a ruffled-up hen. Grows at the bases of oaks and other hardwoods in eastern North America, especially New England, but a smaller, cultivated form is also available under the Japanese name, maitake. Both are excellent.

Giant puffballs (*Calvatia gigantea* and relatives) are enormous ball-like mushrooms of prairies and open ground. Sometimes sold by the slice, but the flesh must be pure white to be edible.

A fully ripe Oregon white truffle (*Tuber gibbosum*) is just as aromatic as its famous European counterpart but is a bargain at less than one-tenth the price. The problem is that these underground fungi are hunted with rakes instead of being sniffed out by dogs or pigs as is the traditional method in Europe. The result is that many are picked before their full aroma and flavour develops.

Oregon black truffles (*Leucangium carthusiana*) are a different fungus altogether from the European black truffle, softer and fruitier with a shorter shelf life. Both Oregon truffles are associated with young Douglas-fir. Most of the harvest takes place on tree farms in the winter and spring.

Cultivated mushrooms such as the shiitake (shiang-gu), portabello, enoki and oyster mushroom are often advertised in restaurants as 'wild mushrooms', which they are not. A better label for them is 'exotic' or 'gourmet' mushrooms, catch-all phrases for virtually any mushroom besides the common supermarket variety.

Sustainability of harvests

Some people have wondered whether the mushroom harvest is sustainable at current levels. While this is a complex subject beyond the scope of this chapter, studies show that as long as the ground isn't dug up deeply, intensive picking has little influence on future crops and may even have a slight stimulatory effect. This is not terribly surprising because the commercially valuable species tend to be more plentiful in second-growth forests or those influenced to some degree by human beings (which is probably why we came to value them in the first place). In Italy, where competition for porcini has been fierce for hundreds of years, they had a bumper crop in 1998.

The good news, then, is that wild mushrooms, like pine nuts or huckleberries, can be harvested without significantly damaging our forests, and that buying them in a gourmet restaurant probably comes at less cost to biodiversity than wine, beef, or almost any other item on the menu. The bad news is that overly restrictive policy and unrealistic permit fees imposed by the US Forest Service and other agencies are turning an already marginal existence into a well nigh impossible one. They make mushroom picking profitable only as an occasional opportunistic activity that does not encourage lifestyle conservation or land stewardship. Such policies portend a future in which packets of dried mushrooms labelled 'Buy Wild Mushrooms – Help Conserve Rain Forests' refer only to forests that are not our own.

Acknowledgement

This chapter is excerpted from a longer piece on the wild mushroom harvest in America published by the author in *California Wild* in the autumn of 1999 and I thank Maurice Rotheroe for his help with this adaptation.

Reference

Gould, S. J. (1997). *Dinosaur in a Haystack*. Penguin Books: Harmondsworth, Middlesex.

9

Threats to biodiversity caused by traditional mushroom cultivation technology in China

SIU WAI CHIU & DAVID MOORE

Introduction

China is the world's major mushroom producing country. *Agaricus* production is mainly for export but *Lentinula edodes* (shiitake or shiang-gu) is the traditional local product, and now the major crop. *Lentinula edodes* is indigenous to China. It was first cultivated there more than 800 years ago, and today, China accounts for about 70% of world production. In 1997, Chinese production was recorded as 91 500 metric tonnes of the dried crop (drying produces the characteristic taste of the mushroom), ten times that in fresh weight. Shiang-gu (the Chinese name) is presently about the second or third most popular cultivated mushroom in the world, being consumed throughout China, Taiwan, Japan and Korea, and with increasing world-wide popularity. One-third of the Chinese crop is exported. As this amounts to the equivalent of about 300 000 tonnes of fresh mushrooms, the industry is an important earner of foreign exchange as well as making a very significant contribution to the income of peasant-farmers especially of the mountainous regions in China (Chang & Chiu, 1992). In these regions the land is poor in fertility and too distant from reliable transport to make conventional farming of green crops profitable.

Traditional technology

The traditional log-pile cultivation method is still the one that is most frequently used. For this, locally felled logs (oak, chestnut, hornbeam, maple and other trees) over 10 cm diameter (probably about 20 to 30 years old) and 1.5 m to 2 m long are normally cut in spring or autumn of each year. Felling at this time minimises pre-infestation by wild fungi or insects. Holes drilled in the logs (or saw- or axe-cuts) are packed with spawn, and the spawn-filled holes are then sealed with wax or other sealant to protect the spawn from the weather. The logs are stacked in laying yards on the open hillside, or under shelters built of bamboo or other materials, in arrangements which permit good air circulation and easy drainage and temperatures between 24 °C and 28 °C. The logs remain here for five to eight months for the fungus to grow completely through the log before transfer to the raising yard to promote fruit body formation. This is usually done in winter to ensure the lower temperature (12–20 °C) and increased moisture that are required for fruit body initiation. The first crops of mushrooms appear in the first spring after being moved to the raising yard. Each log will produce 0.5 kg to 3 kg of mushrooms, each spring and autumn, for 5 to 7 years. We now have enough information for a 'back of an envelope calculation': if each log produces 5 kg mushrooms per year; total annual production today is probably in excess of one million metric tonnes so we need the equivalent of 200 000 2-m logs. This amounts to between 50 000 and 100 000 trees every year just to maintain current production levels. Traditional usage of natural wood logs has been pursued to the extent that as availability of mature trees has declined attention has turned to younger trees and other tree species. This, combined with other demands for timber and land, has contributed to a loss of 87% of the native forests in China (Anon., 1997). China now faces the problem that the rate of deforestation is much greater than the rate of reforestation in the remaining 13% forest cover (Mackinnon *et al.*, 1996; Loh *et al.*, 1999)! There are regulations about planting and prohibitions on felling young trees, but these are difficult to monitor and 'conservation awareness' is especially low among those poor peasants living in the remote mountainous regions.

Conservation issues

The conservation pressures of *Lentinula* production result from the scale of the industry and from the biological consequences of traditional practices. There has been a minimum four-fold increase in production of *Lentinula* over the past twenty years or so, and every prospect that demand for the crop will continue to increase. The conservation pressure the industry exerts is more likely to increase than to ease. The scale is difficult for Europeans to imagine. The crop is still grown mostly by open outdoor cultivation on a very large number of small farms. One well-informed commentator (Luo, 1998) has claimed that there are *10 million mushroom farmers* in China! These mushroom farms are distributed over the whole of the central highlands in China. This means that the *Lentinula* growing region covers *an area about equal to the entire land area of the European Union.*

The main conservation issues raised by the traditional cultivation practices are:

- the obvious impact on wooded hillsides caused by felling of mature trees (of which only the largest logs are used);
- the fact that outside cultivation is used *and the mushroom crop is harvested after basidiospore release has started* with the consequent (unknown) danger(s) resulting from cross contamination between cultivated and natural populations of *Lentinula edodes.*

Population biology

Over the past several years we have investigated the cultivation physiology of *Lentinula edodes* with a view to providing a good scientific basis for using alternative cultivation substrates (Tan & Moore, 1992, 1995), but we have put greater efforts into studying the population biology of *Lentinula edodes* by examining both the cultivated (Chiu *et al.*, 1996) and natural populations (Chiu *et al.*, 1998*a*,*b*, 1999*a*,*b*). The research covers a geographical area which is around 1700 km north to south and 700 km west to east (Fig. 9.1), but includes detailed surveys down to individual logs, and deals with phenotypes varying from morphology and palatability to DNA sequences.

Nineteen strains of *Lentinula edodes* which are used for spawn production for farms throughout mainland China were characterised with three arbitrarily-primed polymerase chain reaction (AP-PCR) profiles, seven random amplified polymorphic DNA markers (RAPD) profiles and five restriction patterns (restriction fragment length polymorphism patterns of the PCR-amplified ribosomal DNAs; rDNA-RFLPs). For AP-PCR, 4–14 DNA bands were amplified for a particular strain while 1–9 DNA bands were amplified using RAPD. Among them only three of the strains tested showed different amplification profiles with most of the primers used. The others showed small differences in three or fewer DNA amplification profiles. All strains showed identical rDNA-RFLPs.

This study indicates that cultivated strains of shiang-gu in China are genetically very homogeneous. In part this probably results from the concentrated artificial breeding work that has been done using a limited range of imported Japanese cultivars, originating in a series of breeding experiments with wild isolates collected from China, Taiwan and New Guinea (Mori, Fukai & Zennyoji, 1974). However, genetic homogeneity is also evident in cultivated strains of *Agaricus bisporus* (Loftus, Moore & Elliott, 1988) and *Volvariella volvacea* (Chiu, Chen & Chang, 1995). Loftus *et al.* (1988) could find no RFLP polymorphisms between three commercial cultivars of *Agaricus bisporus*. They noted that the growth conditions and flushing (times of fruiting) patterns for each of the three strains were also very similar, despite the fact that the three strains were marketed by different companies as original and independent products. These authors commented on how remarkable was ' . . . the coincidence of their genetic similarity . . . '. Since the same remarkable genetic similarity between allegedly different commercial cultivars has been encountered in *Volvariella volvacea* (Chiu *et al.*, 1995) and now in *Lentinula edodes*, it is entirely feasible that genetic homogeneity in cultivated mushrooms does not result from any peculiarity of mushroom genetics but most probably from behaviour patterns of mushroom growers around the world.

Although a very narrow gene pool is used in cultivated strains, our survey of diversity of rDNA sequences indicates that China harbours the greatest germplasm resource of the mushroom *Lentinula edodes* (Chiu *et al.*, 2000). The internal transcribed spacer (ITS) regions of rDNA in *Lentinula edodes* are rather conserved and therefore can be used to trace lineage relationships. Hibbett *et al.* (1995) and Hibbett, Hansen & Donoghue (1998) have identified five lineages in Asia–Australasia with two lineages appearing in China. As the sample size of all previous studies is rather small, we carried out a large-scale and more detailed screening in China, focusing attention on three provinces that are not among those in which traditional cultivation is popular (to avoid the danger of contamination with the 'commercial gene pool'). These are central Hubei, north central Shaanxi and the most Southwest Yunnan (Fig. 9.1). Several fruit bodies from single fallen logs and fruit bodies from different logs were collected and used for culture isolation. Biomass grown in pure culture was used for small scale DNA 'mini-preparations' (Chiu *et al.*, 1996), followed

---

Fig. 9.1. (*opposite*) The experimental area. The three provinces from which field isolates were collected (Shaanxi, Hubei and Yunnan, the latter sharing borders with Myanmar, Laos and Vietnam) are shown shaded in the upper map. In the larger scale map in the lower part of the figure the location of traditional shiang-gu cultivation provinces, Guangdong, Fujian and Jiangxi (and provincial capital cities), is indicated. Commercial cultivars were collected from institutes/universities in Beijing, Hubei, Shanghai, Guangdong, Fujian and Hong Kong.

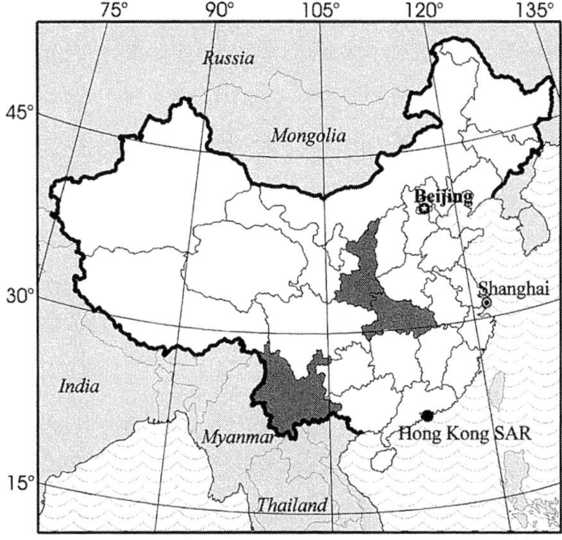

Modified from Magellan Geographix Map in CorelDRAW 8

by specific polymerase chain reaction using a fungal-specific primer set (ITS4 and ITS5). Direct sequencing was performed with the purified PCR-amplified rDNA fragments.

The results reveal that a dominant group I rDNA lineage appears in these three provinces. Compared with the rest, the Group I lineage has the widest geographical distribution, extending from North Korea and Japan to southern China and Thailand. This could indicate an ancient history for this lineage but as many Japanese and Chinese cultivars belong to the Group I lineage it is more probably the effect of breeding work and subsequent adoption of those cultivars for commercial cultivation in these regions. Similarly, wild isolates found to show group I rDNA sequences that were collected in the traditional shiang-gu growing areas of Fujian and Guangdong provinces may indicate cross contamination of the wild population with cultivar genes or even escape of the cultivar to the wild. The collection made by the authors in remote areas such as reserve areas of national status or mountainous areas far beyond human access could answer this problem. In the three provinces Hubei, Shaanxi and Yunnan, group I rDNA lineage sequences were found in the wild isolates. Thus, group I rDNA lineage occupies the largest territory and China might be the home for this lineage. Isolates of the group IV rDNA lineage previously reported in Nepal and Hubei were found in Shaanxi province instead. However, novel rDNA sequences (not belonging to either of the five lineages previously described) were found in the three provinces. Thus, several additional new rDNA lineages are present in China. Besides, as isolates of different rDNA lineages appeared in one single log, it is evident that genetic recombination is possible among the different lineage populations existing in the remote mountainous areas of China.

To investigate small-scale population structure, a field study was carried out in a remote broad-leaved *Fagus longipetiolata* forest in Shaanxi province. Following spatial mapping, 24 fruit bodies were collected for tissue isolation into axenic culture. Twenty-four genets distributed on fallen tree trunks within a distance of 120 m were identified and clustered into 7 groups using the unweighted pair–group method algorithm using data based on colony morphologies, abilities to degrade aromatic poly-R478 dye, somatic incompatibility reaction patterns and DNA fingerprints. Among the parameters used, the somatic incompatibility reaction, a polygenic phenotype, was the most differentiating, identifying 22 incompatible classes. Two sets of fruit bodies of different genets were so close together that they would otherwise have been described as aggregate fruits of presumed identical origin. Eighteen genets found on the same 5.6 m long tree trunk divided roughly into two clusters, matching their spatial distribution, and a nearby branch bore another distinct cluster. More heterogeneity was encountered between isolates the greater the distance separating them on the original site. Genets on the same tree trunk showed more compatible somatic reactions among themselves, and their DNA fingerprints showed higher similarity. Nevertheless, considering the totality of phenotypic characters, each fruit body is a genet in *L. edodes*.

Such features are concluded to result from a reproductive strategy that depends on basidiospore dispersal. Within each cluster of isolates from the collection site genets seemed to have arisen from multiple sib-mating events. Thus, a cluster may represent a lineage of *L. edodes*. Individualism in *L. edodes* is based on a strong somatic incompatibility system. Strong competition from contaminating individuals arriving as air-borne basidiospores could explain decreased and fluctuating crop yields which are now frequently observed in later flushes from the outdoor wood log cultivation system. Further, it would also explain why multispore spawn is not favoured in artificial cultivation of this economically important edible mushroom. This also applies to *Agaricus* and *Pleurotus* mushrooms which, together with *Lentinula* form the triumvirate of the most popular cultivated species in the world, with China being the main producer and exporter.

Improved technology

The traditional approach to shiang-gu production is demanding and exploitative in its use of both land and trees. For these reasons more industrial approaches are being applied to shiang-gu growing. Hardwood chips, sawdust and other solid agricultural wastes packed into polythene bags as 'artificial logs' provide a highly productive alternative to the traditional technique, and the cultivation can be done in houses (which may only be plastic-covered enclosures) in which climate control allows year-round production. The industry raises other conservation issues that our recent analyses of molecular and conventional genetic markers have addressed. A very limited gene pool is exploited in the cultivated strains in China, yet there is an enormous biodiversity in the species in the wild. Analysis of local populations reveals that *L. edodes* strains show strong somatic incompatibility reactions and individual territories can be small (a few hundred mm). The widespread nature of the species and absence of other means of dispersal indicate that basidiospores are the major, even only, method of natural distribution. As the crop is harvested *after* initiation of basidiospore release and the crop is mostly grown outdoors, cross contamination between wild and cultivated strains is inevitable. This puts the natural gene pool under threat. Protection of the natural environment is still the best strategy for conserving the biodiversity of this important commercial resource. Collection of wild strains for preservation in a culture collection would directly conserve the wild germplasm. Making such a gene bank readily accessible to the public and industry would also generate a commercial resource for exploitation in both cultivation and breeding programmes. It may also reduce the pressure caused by non-professional collection from nature.

Our studies suggest that a move to indoor cultivation, less dependence on multispore spawns and exploitation of a wider range of natural genotypes would better safeguard both cultivated and natural populations of the fungus and avoid denuding hillsides of mature trees. The concept of sustainable management is too novel for the mushroom farmers of a third world country. But villages in China are accustomed to working co-operatively to establish a shared facility, and the advantages of indoor cultivation to the farmer (consistency of yield, much shorter production cycle, use of solid industrial/agricultural wastes in the substrate) can be readily appreciated. An essential step is to educate the public to introduce these ideas. Education is the key.

References

Anon. (1997). *The Chinese Forests*. Forestry Department: Beijing, People's Republic of China.

Chang, S. T. & Chiu, S. W. (1992). Mushroom production – an economic measure in maintenance of food security. In *Microbial Technology: Economical and Social Aspects* (ed. E. J. DaSilva, C. Ratledge & A. Sasson), pp. 110–141. Cambridge University Press: Cambridge, UK.

Chiu, S. W., Chen, M. J. & Chang, S. T. (1995). Differentiating homothallic *Volvariella* mushrooms by RFLPs and AP-PCR. *Mycological Research* **99**, 333–336.

Chiu, S. W., Chiu, W. T., Lin, F. C. & Moore, D. (2000). Diversity of rDNA sequences indicates that China harbours the greatest germplasm resource of the cultivated mushroom *Lentinula edodes*. In *Science and Cultivation of Edible Fungi*, vol. 1 (ed. L. J. L. D. Van Griensven), pp. 239–243. A. A. Balkema: Brook Field.

Chiu, S. W., Ma, A. M., Lin, F. C. & Moore, D. (1996). Genetic homogeneity of cultivated strains of shiitake (*Lentinula edodes*) used in China as revealed by the polymerase chain reaction. *Mycological Research* **100**, 1393–1399.

Chiu, S. W., Wang, Z. M., Chiu, W. T., Lin, F. C. & Moore, D. (1999*a*). An integrated study of individualism in *Lentinula edodes* in nature and its implication for cultivation strategy. *Mycological Research* **103**, 651–660.

Chiu, S. W., Yip, M. L., Leung, T. M., Wang, Z. W., Lin, F. C. & Moore, D. (1998*a*). A preliminary survey of genetic diversity in a natural population of shiitake (*Lentinula edodes*) in China. Proceedings, Fourth Conference on the Genetics and Cell Biology of Basidiomycetes, Nijmegen, The Netherlands, 27–30 March, 1998, p. 175. The Mushroom Experimental Station: Horst, The Netherlands.

Chiu, S. W., Yip, M. L., Leung, T. M., Wang, Z. W., Lin, F. C. & Moore, D. (1998*b*). Genetic diversity in a natural population of shiitake (*Lentinula edodes*) in China. Abstracts of the Sixth International Mycological Congress, Jerusalem, Israel, 23–28 August 1998, p. 34.

Chiu, S. W., Yip, P. & Moore, D. (1999*b*). Genetic diversity of *Lentinula edodes* collected from Hubei Province, China. Poster in Abstracts, Seventh International Fungal Biology Conference (ed. J. H. Sietsma), University of Groningen, The Netherlands, 22–25 August, 1999.

Hibbett, D. S., Fukumasa-Nakai, Y., Tsuneda, A. & Donoghue, M. J. (1995). Phylogenetic diversity in shiitake inferred from nuclear ribosomal DNA sequence. *Mycologia* **87**, 618–638.

Hibbett, D. S., Hansen, K. & Donoghue, M. J. (1998). Phylogeny and biogeography of *Lentinula* inferred from an expanded rDNA dataset. *Mycological Research* **102**, 1041–1049.

Loftus, M. G., Moore, D. & Elliott, T. J. (1988). DNA polymorphism in commercial and wild strains of the cultivated mushroom, *Agaricus bisporus*. *Theoretical and Applied Genetics* **76**, 712–718.

Loh, J., Randers, J., MacGillivray, A., Kapos, V., Jenkins, M., Groombridge, B., Cox, N. & Warren, B. (1999). *The Living Planet*. WWF International: Gland.

Luo, X.-C. (1998). Mushroom Genetic Resource, Evaluation and Utilization in China. In *Proceedings of the '98 Nanjing International Symposium, Science and Cultivation of Mushrooms* (ed. M. Lu, K. Gao, H.-F. Si & M.-J. Chen), pp. 142–152. JSTC-ISMS: Nanjing, China.

Mackinnon, J., Sha, M., Cheung, C., Carey, G., Zhu, X. & Melville, D. (1996). *A Biodiversity Review of China*. WWF International: Hong Kong.

Mori, K., Fukai, S. & Zennyoji, A. (1974). Hybridization of shiitake (*Lentinus edodes*) between cultivated strains of Japan and wild strains grown in Taiwan and New Guinea. *Mushroom Science* **9**, 391–403.

Tan, Y.-H. & Moore, D. (1992). Convenient and effective methods for *in vitro* cultivation of mycelium and fruiting bodies of *Lentinus edodes*. *Mycological Research* **96**, 1077–1084.

Tan, Y.-H. & Moore, D. (1995). Glucose catabolic pathways in *Lentinula edodes* determined with radiorespirometry and enzymic analysis. *Mycological Research* **99**, 859–866.

10

A preliminary survey of waxcap grassland indicator species in South Wales

MAURICE ROTHEROE

Introduction

Over the past several decades conservationists have become increasingly concerned about the disappearance of traditionally managed meadows and pastures. These grasslands and their characteristic communities of native plants have been lost through treatment with artificial fertilisers and selective herbicides, through ploughing and reseeding and by encroaching development. It was against this background in 1987 that the Countryside Council for Wales (CCW) began a systematic survey of semi-natural grasslands in Wales (currently nearing completion), using the National Vegetation Classification (NVC) (Rodwell, 1991, *et seq.*) to define plant communities. As well as providing basic information on the distribution and extent of unimproved grassland, a further aim of this CCW survey was to identify the best examples of this habitat in Wales and to afford them official recognition and protection as Sites of Special Scientific Interest (SSSIs). These are areas of land or water which are designated as being of nature conservation importance for their biological, geological or physiographic features. Designation affords the site statutory protection and aims to secure its sympathetic management. Currently, there are more than 1000 SSSIs in Wales.

In parallel with vascular plants, a series of species of macrofungi is largely confined to, and therefore indicative of, traditionally managed grasslands. Such habitats are known to mycologists as *Hygrocybe* grasslands (Feehan & McHugh, 1992) or waxcap grasslands (Rotheroe *et al.*, 1996). The genus *Hygrocybe* (waxcaps) is not the only fungal indicator of this grassland community. Waxcaps are often accompanied by fairy clubs (clavarioid fungi); pink-spored species of the genus *Entoloma* (pink-gills); and earth-tongues (family Geoglossaceae). Many of the taxa in these groups feature in the British and European Red Data lists of threatened or endangered fungi (Ing, 1992, 1993). Several are also listed in many of the national Red Data lists in Europe. For example, Arnolds & de Vries (1993) have stated that all European species of *Clavaria* (fairy clubs) are included on a Red Data list somewhere in Europe, while 67% of another grassland fairy club genus, *Clavulinopsis*, feature on at least one European Red Data list. Figures for other genera they gave are: *Hygrocybe*, 89% and *Entoloma sensu lato*, 97%.

Criteria for the selection and notification of biological SSSIs largely ignore the fungal element of the community, principally because of the absence of data on species distribution and abundance (Hodgetts, 1992). This has led to concern that mycologically rich sites may be under-represented within the SSSI series, a situation that may have been exacerbated by the fact that certain waxcap grasslands can be floristically impoverished (Keizer, 1993). Responding to this data deficiency, the British Mycological Society (BMS) initiated a survey throughout the British Isles of the fungi characterising unimproved meadow and pasture grassland (Rotheroe *et al.*, 1996). This began in 1996 and is an ongoing survey.

The main aim of the South Wales project was to record fruiting of waxcap grassland indicator species at a number of grassland sites in Carmarthenshire, a county somewhat under-recorded mycologically. Most of the survey sites were selected on the basis of their rich vascular plant flora. The additional sites, both inside and outside the county, were included for their known mycological interest and for comparative purposes. The survey also aimed to examine any relationship between mycological interest and NVC type or management regime. The survey was commissioned by CCW and this chapter represents a summary of the report to that agency (Rotheroe, 1999).

Site selection

Ten target sites were identified in conjunction with CCW officers. The selected locations embraced a number of different NVC types, including examples of Mesotrophic Grassland (symbolised MG); Mire (M); Calcifugous Grassland (U); and Calcicolous Grassland (CG). These sites are listed in Table 10.1. Visits were also made to seven additional sites (listed in Table 10.2), but NVC surveys have not been carried out at these additional locations.

Past records from sites 1, 9, 11, 13 and 17 enabled cumulative lists to be compiled. Cumulative data were also obtained from recording visits by

Table 10.1. *Details of the ten Carmarthenshire sites selected for survey*

Site number	Location	Code	Grid reference	NVC types represented (data supplied by CCW)
1	'Talley' SSSI[a]	TAL	SN6–3–	MG5a, MG5c, M25b
2	Pwll Edrychiad SSSI	PED	SN584162	MG5a, MG5c, M25b[b]
3	Caeau Nant Garenig SSSI	CNG	SN673124	MG5, M25b, M24b/c[c]
4	Cae Blaen-dyffryn SSSI[d]	CBD	SN604445	MG5a[c], MG6b (in annex)
5	Caeau Blaen-bydernyn SSSI	CBB	SN558439	U4c, M25b
6	Carreg Cennen SSSI	CCN	SN670191	CG1e, CG2c, MG6b
7	Whitehill Down SSSI	WHD	SN290135	MG5a, M24b/c[c]
8	Rhosydd Castell-du SSSI	RCD	SN655116	MG5c, M24b/c, M25b
9	Waun-las (Middleton)	WLS	SN528178	MG6b
10	Caeau Caradog	CCG	SN694460	MG5a, MG5c, U4a

[a]The owner of this site requested anonymity. [b]Also includes some secondary calcareous grassland around rocky outcrops. [c]Also includes small stands of MG5c. [d]Also includes Annex to SSSI.
Key to grassland communities and subcommunities (for fuller details see Rodwell, 1991 *et seq.*): Calcicolous Grassland, CG1 & CG2 (*Festuca ovina–Carlina vulgaris* grassland and *Festuca ovina–Avenula pratensis* grassland). Both are listed above as CG. Mires, M23a and M25b (*Juncus effusus/acutiflorus–Galium palustre* rush–pasture: *Juncus acutiflorus* subcommunity and *Molinia caerulea–Potentilla erecta* mire: *Anthoxanthum odoratum* subcommunity). Mesotrophic Grassland, MG5a and MC5c (*Cynosurus cristatus–Centaurea nigra* grassland: *Lathyrus pratensis* subcommunity and *Danthonia decumbens* subcommunity). Mesotrophic Grassland, MG6 (*Lolium perenne–Cynosurus cristatus* grassland). Calcifugous Grassland, U4a; U4b & U4c (*Festuca ovina–Agrostis capillaris–Galium saxatile* grassland typical subcommunity, *Holcus lanatus–Trifolium repens* subcommunity and *Lathyrus montanus–Stachys betonica* subcommunity).

other workers to Gilfach Farm, Rhyader, Radnorshire (Ray Woods, pers. comm.). This site is abbreviated by the code letters GILF.

Management regimes

The management regimes that were in place are listed in Table 10.3 as far as they are known. This information has been obtained from CCW, the site owners and from personal observations.

Table 10.2. *Additional sites surveyed*

Site number	Location	Code	Grid reference
11	Llanerchaeron Estate (NT), Cardiganshire	LLAN	SN480600
12	Llanrhystud Chapelyard, Cardiganshire	LRST	SN547692
13	Hafod Estate, Cardiganshire	HAFD	SN756731
14	Dinefwr Park (NT), Carmarthenshire	DINF	SN617225
15	Maestir Churchyard, Lampeter, Cardiganshire	MAES	SN553494
16	St David's College lawns, Lampeter, Cardiganshire	STDC	SN579482
17	Garn Ddyrys, Tumble, Breconshire	GARN	SO258117

Field procedures

Each of the 17 sites was visited at least twice and some sites were visited more often. Each recording visit lasted between two and four hours, depending upon the size of the site. Recording effort was concentrated in relevant stands of vegetation which were identified from CCW's Phase II vegetation maps. Once a homogeneous stand representing one of the target NVC types was located, a 'mowing' transect was carried out, by walking back and forth across the area as if mowing a lawn. However, in fields and areas which from the outset could be regarded as lacking any fungal fruiting (usually the mire sites), a less rigorous system of transects was used. Relevant NVC type and management regimes, from Phase II reports and from field observations, were also noted. Fruit bodies of common or easily identifiable fungi were left *in situ* but for many identifications it was necessary to collect representative samples for closer macroscopic and microscopic examination in the laboratory. Dried voucher material was retained for all but the commonest species and this material was deposited in the herbarium of the National Botanic Garden of Wales (NBGW) or in Herb K(M) (Royal Botanic Gardens, Kew). The data obtained from all the fieldwork were entered into the BMS database.

Evaluation of conservation value

Conservation of this unimproved grassland habitat has been the subject of intensive study in recent years. The results of these studies provided a well-documented context in which to evaluate the findings of this survey.

The data generated in the South Wales survey were analysed using the

Table 10.3. *Management history at survey sites*

Site number	Location and vegetation	Management history
1	'Talley' SSSI (MG5a, MG5c, M25b)	Pasture plus two hay meadows, cattle-grazed. Farmed by organic methods.
2	Pwll Edrychiad SSSI (MG5a, MG5c, M25b)	Pasture, cattle and sheep-grazed, with clear evidence of rabbit grazing. No evidence of fertiliser treatment.
3	Caeau Nant Garenig SSSI (MG5, M25b, M24b/c)	Pasture and some *Molinia*-dominated areas cut for hay. Sheep and pony grazing.
4	Cae Blaen-dyffryn SSSI and Annex (MG5a)	Pasture, sheep-grazed. Unfertilised (Annex outside SSSI had some recent treatment).
5	Caeau Blaen-bydernyn SSSI (U4c, M25b)	Pasture, cattle and sheep-grazed. Unfertilised.
6	Carreg Cennen SSSI (CG1e, CG2c, MG6b)	Pasture, cattle and sheep-grazed. Neutral grassland 'heavily modified' according to SSSI notification documentation.
7	Whitehill Down SSSI (MG5a, M24b/c)	Hay cutting with aftermath grazing by cattle. Artificial fertiliser used in certain parts. Some areas show past disturbance 'probably ploughing'.
8	Rhosydd Castell-du a Plas y Bettws SSSI (MG5c, M24b/c, M25b)	Pasture, cattle-grazed. No evidence of fertiliser treatment.
9	Waun-las (Middleton) (MG6b)	Pasture, cattle and sheep-grazed, but sheep withdrawn from large areas for 12 months prior to survey. Previous owners claim no artificial fertiliser or ploughing in living memory.
10	Caeau Caradog (MG5a, MG5c, U4a)	Pasture, horse, pony and sheep-grazed, very heavily in some areas. MG5a has suffered substantial modification.
11	Llanerchaeron Estate (NT) (MG5?)	Lawn, grazed by rabbits and occasionally sheep escaping from adjoining fields. Mown twice a year. No fertiliser or treatment within living memory.
12	Llanrhystud Chapelyard (?)	Ungrazed, mown twice a year. No fertiliser.
13	Hafod Estate (?)	Pasture, relatively heavy sheep-grazing. No fertiliser or treatment in living memory.
14	Dinefwr Park (NT) (?)	Deer park, deer and cattle-grazed. Owners claim no fertiliser treatment, but some past treatment is possible.
15	Maestir Churchyard, Lampeter, Cardiganshire (?)	Ungrazed, cut 'regularly'. Believed no fertiliser.
16	St David's College lawns, Lampeter (?)	Ungrazed, cut 'regularly'. Believed no fertiliser.
17	Garn Ddyrys, Tumble (?)	Pasture, sheep and rabbit-grazed, relatively heavily in many areas. No information on fertiliser treatment, but this is unlikely to have taken place.

Table 10.4. *Separation of conservation values into four classes according to Rald (1985)*

Conservation value	Number of *Hygrocybe* species recorded on a single visit	Total number of *Hygrocybe* species expected to be recorded
National importance (I)	11–20	17–32
Regional importance (II)	6–10	9–16
Local importance (III)	3–5	4–8
No importance (IV)	1–2	1–3

methodologies proposed by Rald (1985) and Rotheroe *et al.* (1996), Rotheroe (1997, 1999). The Rald formula is based on the cumulative number of *Hygrocybe* species recorded at a location, with numbers recorded on a single visit enabling tentative predictions of the full mycota to be made. Rald separated the conservation values into four classes (Table 10.4).

Rotheroe (1999) proposed that an indicator-species profile of a site would refine the evaluation of conservation value. This was a modified version of various weighting systems elaborated by a number of Scandinavian workers (John Bjarne Jordal, pers. comm.). The profile can be expressed in a shorthand using the following code letters: **C** (clavarioid fungi – fairy clubs); **H** (*Hygrocybe sensu lato* – waxcaps); **E** (grassland species of the Entolomataceae – pink-gills); **G** (Geoglossaceae – earth-tongues). Thus a site with survey records of eight waxcaps, two fairy clubs, two earth-tongues and one species of *Leptonia*, would be described as: **C**2, **H**8, **E**1, **G**2. This shorthand system has been used in the present study. In the interests of simplicity, *Dermoloma* and *Porpoloma* are merged with **H**. This is legitimate, since they have the same ecological affinities and *Porpoloma* was, until recent taxonomic revisions, traditionally included in the genus *Hygrophorus* (as was *Hygrocybe*), while the current classification places *Dermoloma* in the tribus Hygrocybeae. The author therefore interpreted Rald in this fashion and also referred to the '**CHEG** profile' as a means of making easy quantitative assessments for comparison of different sites and to suggest their relative conservation value. In using these numerical formulae a variety record is given equal weighting to that of a species. Thus the numbers refer to taxa, rather than to species.

Table 10.5. *Highest number of* Hygrocybe *(waxcap) taxa recorded on single visits made in 1998*

		Conservation value				
		Indicator species profile[a]				
Site number	Location	C	H	E	G	Assigned according to Rald (1985)[b]
17	GARN	1	17	0	4	National importance
13	HAFD	1	15	0	0	National importance
6	CCN	3	14	1	0	National importance
11	LLAN	4	13	4	1	National importance
9	WLS	2	10	3	0	Regional importance
1	TAL	1	9	6	0	Regional importance
16	STDC	2	9	3	3	Regional importance
2	PED	1	8	3	1	Regional importance
15	MAES	2	7	1	2	Regional importance
4	CBD	0	6	2	0	Regional importance
14	DINF	1	6	0	0	Regional importance
5	CBB	1	4	0	0	Local importance
12	LRST	0	4	0	0	Local importance
7	WHD	1	3	0	0	Local importance
10	CCG	0	3	0	0	Local importance
3	CNG	0	0	0	0	
8	RCD	0	0	0	0	

[a]Rotheroe (1999) and see text. [b]See Table 10.4.

Nomenclature

In general the classification of the *Dictionary of the Fungi* (Hawksworth *et al.*, 1995) has been used in this study. However, there is some disagreement currently over aspects of nomenclature for certain groups. For the Entolomataceae (pink-gill species), this study uses the nomenclature of Noordeloos (1992), which reduces some genera to subgeneric level, notably *Leptonia* and *Nolanea*, rather than the classification of the *Dictionary of the Fungi*, which retains these at generic level. For *Hygrocybe*, the nomenclature follows Boertmann (1995), with a number of small modifications proposed by Henrici (1996).

Analysis of results

Table 10.5 shows the total number of waxcap species recorded on a single visit to survey sites, and Table 10.6 the cumulative totals of indicator

Table 10.6. *Cumulative totals of indicator species recorded to the end of December 1998*

		Conservation value				
		Indicator species profile[a]				
Site number	Location	C	H	E	G	Assigned according to Rald (1985)[b]
17	GARN	3	28	0	5	National importance
13	HAFD	4	26	7	0	National importance
11	LLAN	6	23	7	2	National importance
	GILF[c]	3	23	0	0	National importance
9	WLS	4	22	7	0	National importance
6	CCN[d]	4	19	4	0	National importance
1	TAL	2	16	14	0	Regional importance
2	PED[d]	3	11	3	1	Regional importance
16	STDC[d]	2	9	3	3	Regional importance
14	DINF[d]	1	9	0	0	Regional importance
15	MAES[d]	2	8	1	2	Local importance
5	CBB[d]	1	6	0	0	Local importance
4	CBD[d]	0	6	2	0	Local importance
12	LRST[d]	0	4	0	0	Local importance
7	WHD[d]	1	3	0	0	
10	CCG[d]	0	3	0	0	
3	CNG[d]	0	0	0	0	
8	RCD[d]	0	0	0	0	

[a]Rotheroe (1999) and see text. [b]See Table 10.4. [c]Gilfach Farm, Rhayader, Radnorshire. [d]Sites for which no previous records existed, so the cumulative totals shown are for 1998 only.

species recorded up to the end of December, 1998. The conservation values shown in these Tables were conceived in a European context and are effectively being extrapolated to Wales. This seems to be justified, since similar analyses in the UK have been applied during the BMS waxcap grassland survey and the results do appear to reflect the conservation importance of the respective sites examined (personal observations).

Correlations with NVC types

Table 10.7 summarises the number of waxcap grassland indicator species recorded from the various NVC grassland types at the sites surveyed. The full list of species is shown in Table 10.8. No NVC surveys had been carried out at the additional sites listed, nor at Waun-las Farm, but it was suggested that most of the grassland at this latter location is referable to

Table 10.7. *Numbers of indicator species recorded at different grassland types categorised using the National Vegetation Classification (NVC)*

NVC grassland type	Number of species recorded
Calcicolous grassland CG	10
Calcifugous grassland U4a	2
Calcifugous grassland U4b	4
Calcifugous grassland U4c	7
Mesotrophic grassland MG5a	19
Mesotrophic grassland MG5c	18
Mesotrophic grassland MG6	34
Mire M23a	2
Mire M24	0
Mire M25b	1

MG6 (Jamie Bevan, pers. comm.). Records from this site are therefore listed under that heading. NVC types for the other additional sites are omitted.

The greatest numbers of species were clearly recorded from the neutral grassland types MG5 and MG6. Fewer species were recorded from calcicolous and calcifugous grassland, although to some extent this may simply reflect under-representation of these habitats within the survey sites. The mire types are clearly very impoverished from a mycological perspective.

Rare, threatened or endangered species

A list of those species recorded at the study sites which feature in the British (Ing, 1992), European (Ing, 1993), and Welsh (Rotheroe, 1998) Red Data lists is shown in Table 10.9. Also indicated are the three species covered by the UK Biodiversity Action Plan (BAP) (Anon., 1999).

Meteorological considerations

The survey area suffered abnormally heavy precipitation in 1998. Total annual rainfall, as reported from two local meteorological stations, was 28% higher than the norm for the past 30 years (John Wildig & John Powell, pers. comm.). One of these stations is upland, at an elevation of 301 m; the second, 100 km to the south, is almost at sea level. Most of the Carmarthenshire sites lie approximately halfway between the two. Further details are given in Rotheroe (1999).

Table 10.8. *Species of macrofungi recorded from the various grassland types*

Species	Grassland type (NVC)
Clavaria vermicularis	MG5a
Clavulinopsis corniculata	CG, MG5a, MG6
Clavulinopsis fusiformis	MG6
Clavulinopsis helvola	CG, M25b, MG5c, MG6, U4c
Clavulinopsis luteoalba	MG5c
Dermoloma cuneifolium	MG5a, MG6
Entoloma anatinum	MG5c
Entoloma conferendum	MG6, U4b
Entoloma corvinum	M23a
Entoloma cruentatum	MG5c
Entoloma hebes	MG5c
Entoloma infula	MG5c, U4b
Entoloma porphyrophaeum	MG6
Entoloma serrulatum	MG6
Hygrocybe calyptriformis	MG5a, MG5c, MG6
Hygrocybe ceracea	MG5a, MG6, MG6
Hygrocybe chlorophana	CG, M23a, MG5a, MG5c, MG6, U4c
Hygrocybe cinereifolia	MG6
Hygrocybe citrinovirens	MG5a, MG6
Hygrocybe coccinea	MG5a, MG6, U4c
Hygrocybe colemanniana	CG, MG6
Hygrocybe conica	MG5a, MG5c, MG6
Hygrocybe conica v. *chloroides*	MG6
Hygrocybe glutinipes	MG6
Hygrocybe insipida	CG, MG5a, MG5c, MG6
Hygrocybe irrigata	MG5a, MG5c, MG6
Hygrocybe laeta	MG5c, MG6, U4a, U4b, U4c
Hygrocybe nitrata	MG6
Hygrocybe persistens	MG6
Hygrocybe pratensis	MG5a, MG5c, MG6, U4a, U4b, U4c
Hygrocybe pratensis v. *pallida*	MG5a, MG5c
Hygrocybe psittacina	MG5a, MG5c, MG6, U4c
Hygrocybe psittacina v. *perplexa*	MG6
Hygrocybe punicea	MG6
Hygrocybe quieta	CG, MG5a, MG6
Hygrocybe reidii	CG, MG5a, MG5c, MG6, U4c
Hygrocybe russocoriacea	CG, MG5a, MG5c, MG6
Hygrocybe splendidissima	MG6
Hygrocybe virginea	CG, MG5a, MG6
Hygrocybe virginea v. *fuscescens*	CG, MG6
Hygrocybe virginea v. *ochraceopallida*	MG5a, MG6
Hygrocybe vitellina	MG5c
Microglossum olivaceum	CG
Ramariopsis kunzei	MG6

Table 10.9. *Rare, threatened or endangered species recorded at grassland sites in the present study*

Species	Threat category[a]	Location(s) where recorded
Clavaria zollingeri	BRDL, WRDL, BAP	GILF, HAFD, LLAN, MAES
Geoglossum fallax	ERDL	LLAN, STDC
Geoglossum glutinosum	ERDL	STDC
Entoloma bloxamii	BRDL, WRDL	LLAN
Entoloma cruentatum	WRDL	TAL
Hygrocybe calyptriformis	BRDL, WRDL, priority BAP sp.	CBD, GARN, GILF, HAFD, LLAN, LRST, MAES, TAL
Hygrocybe cantharellus	ERDL	GILF
Hygrocybe fornicata	ERDL	GARN, HAFD, STDC, WLS
Hygrocybe insipida	ERDL	CCN, GARN, HAFD, LLAN, PED, TAL, WLS
Hygrocybe intermedia	ERDL	DINF, GILF, WLS
Hygrocybe irrigata	ERDL	GILF, HAFD, LLAN, TAL, WLS
Hygrocybe nitrata	ERDL	GARN, GILF, WLS
Hygrocybe ovina	ERDL	GILF, LLAN
Hygrocybe psittacina v. *perplexa*	ERDL	CCN, HAFD
Hygrocybe punicea	ERDL	CCN, GARN, GILF, HAFD, LLAN, MAES, WLS
Hygrocybe quieta	ERDL, WRDL	CBD, GARN, HAFD, PED
Microglossum olivaceum	BRDL, ERDL, WRDL, priority BAP sp.	GARN, LLAN, MAES, PED
Porpoloma metapodium	BRDL, WRDL	DINF, HAFD, LLAN
Ramariopsis kunzei	WRDL	CCN, HAFD
Trichoglossum hirsutum	ERDL	GARN, MAES
Trichoglossum walteri	WRDL	STDC

[a]BRDL, British Red Data list (Ing, 1992); ERDL, European Red Data list (Ing, 1993); WRDL, Welsh Red Data list (Rotheroe, 1998); BAP, UK Biodiversity Action Plan (Anon., 1999).

All the sites visited were showing the effects of continuous heavy rainfall. This resulted in sodden ground and anaerobic soil conditions which persisted for long periods during what is regarded as the 'normal autumn fungus season'. Because of the waterlogged conditions, sites grazed by cattle or horses suffered additional damage through 'poaching' of the soil, probably disrupting underground mycelial networks. Previous work has shown that mire communities hold only marginal mycological interest (Rotheroe, 1997) and sites supporting these communities were particularly adversely affected by the excessive rain.

Comments on some individual sites

Many of the locations surveyed are already nature reserves or SSSIs, but none has been notified as such because of any mycological consideration. However, all the following sites should, in the author's view, be listed as *Important Areas for Fungi*. This category refers to a new initiative sponsored by the charity Plantlife to compile an inventory of mycologically important sites in Britain. It is an ongoing project. Summarised below are some of the best of the sites covered by the survey.

'Talley' SSSI

This site gains a Regional Importance ranking in Tables 10.5 and 10.6, featuring in sixth and seventh position amongst the top ten sites. However, there are reasons for regarding it as being of National Importance. It is remarkable for the comparatively large number of *Entoloma* species recorded there, the cumulative total being 14, which is double that of any other location in this survey. Among them is *Entoloma cruentatum*, which was only the second British record when it was first collected there in 1996, although is also now known from two other locations in Britain.

Pwll Edrychiad SSSI

A site that appears to be potentially of National Importance, although the Rald formula rates it as being only of Regional Importance. It is the only site in Carmarthenshire where the olivaceous earth-tongue (*Microglossum olivaceum*) has been recorded. This rare fungus is known from only six other localities in Wales. It has a westerly distribution in mainland Britain and Wales appears to be one of its main population centres. It was fruiting at Pwll Edrychiad in October 1998 in large numbers (over 200 fruit bodies in two colonies) on small tracts of open, rocky calcicolous grassland. Two visits probably revealed only a fraction of this SSSI's full mycota.

Carreg Cennen SSSI

Carreg Cennen was included in the survey for its stands of calcicolous grassland (a scarce resource in Carmarthenshire), although these form relatively small, narrow tracts. Paradoxically, however, most of the indicator species were recorded on the surrounding, extensive MG6b grassland areas. This was one of the best sites visited and easily qualifies as being of National Importance, with a cumulative indicator score of **C**4 **H**19 **E**4 **G**0. It seems likely that this site has a far more extensive and rich mycota than has been revealed from a single year's data.

Waun-las

Having been surveyed in the previous two years, Waun-las was already known to have a rich grassland mycota. On the first visit in October 1996, a total of 16 different species of *Hygrocybe* were recorded. It was noted in 1998, however, that grazing had been withdrawn during the preceding 12 months. The grass was therefore longer and ranker than previously. Not only were the number of waxcap species recorded in 1998 reduced, but the number of fruit bodies in evidence showed a considerable drop compared with previous years. Previous grazing by sheep in this area had maintained a relatively short turf, with good bryophyte cover. It is understood that shortly after the final recording visit to the site in 1998, sheep grazing was resumed (Janet Moseley, pers. comm.). The current owners, the National Botanic Garden of Wales, have already made approaches to CCW, suggesting that the mycological richness of these unimproved grasslands warrants their being notified as an SSSI. They acknowledge, however, that the grasslands are unremarkable floristically and appear to be impoverished MG6b *Lolium–Cynosurus* grassland. This, taken together with the observations at Carreg Cennin above, indicate that a rich mycota can persist in partially improved neutral grassland provided that grazing pressures are sufficient to maintain short-cropped swards.

Llanerchaeron Estate

This was an additional site, included because its small unimproved lawn has previously shown it to be one of the best waxcap sites in Wales. It deserves SSSI status, but certainly qualifies as an *Important Area for Fungi.*

Hafod Estate

This estate has a large area of traditionally managed grassland, with rather intensive grazing by sheep. It is not highly regarded for its floristic qualities. However, the CHEG rating confirms its mycological importance. Every year new species of the indicator group are recorded there.

General conclusions

In South Wales, 1998 was an atypical year so far as the autumn fungus season was concerned. Fruiting of macrofungi in grasslands was delayed and rainfall much higher than average resulted in waterlogged conditions at many of the study sites. Nevertheless, some valuable county data were obtained and comparisons could be made with sites known to be mycologically important for grassland indicator species. Four Carmarthenshire sites are judged to be of high conservation value. These are: Waun-las Farm, Middleton; Carreg Cennen SSSI; 'Talley' SSSI; and Pwll Edrychiad SSSI. The first two are regarded as being of putative National Importance and the other two might also qualify for this ranking. Two of the remaining sites, Caeau Caradog and Whitehill Down SSSI, may well have a much richer mycota than was observed in this survey.

Grassland types

The richest grasslands were clearly the mesotrophic types. Plentiful data from MG5 *Cynosurus–Centaurea* grassland was perhaps expected from past experience of this community. However, it may be surprising to higher plant ecologists to discover that some of the most mycologically productive habitats were those of the MG6 *Lolium–Cynosurus* type.

It is clear from this survey that certain less modified forms of MG6 grassland can be rich in fungi despite appearing impoverished from a vascular plant perspective. At both Carreg Cennen and Waun-las a clear distinction can be made from mycological data between MG6 grassland which has been heavily fertilised or improved in some way and the same grassland type which has been less modified. This distinction is often a dramatic one.

This is perhaps a consideration which the users of NVC for conservation evaluation should address. The conservation importance of some MG6 sites may be considerably underestimated unless they are visited by a mycologist during the late summer and autumn fruiting season.

Calcicolous grasslands are poorly represented in Carmarthenshire and the survey results do not reflect the potential of this NVC type. Mires M23, M24 and M25 were mycologically poor.

Management recommendations

At those sites which have been identified as being mycologically important the main recommendation is to maintain the *status quo*, that is to refrain from fertiliser application or other chemical treatments and to continue at existing stocking levels. The mycological interest of some of the remaining sites might be enhanced if fertiliser treatments were withheld and stocking levels increased slightly. If a choice is available, then grazing by sheep, even at quite heavy stocking levels, as at Hafod, appears to be the most successful regime to encourage fungal fruiting, although cattle grazing at relatively low intensity (such as occurs at 'Talley' SSSI) can clearly maintain favourable conditions too. Personal observations suggest that rabbit populations should be encouraged. This grazer produces close-cropped areas, which are extremely productive for certain indicator fungal species.

Future work

Within Carmarthenshire further mycological recording is recommended, particularly at Carreg Cennen SSSI, Pwll Edrychiad SSSI, Caeau Caradog and Whitehill Down SSSI. This should ideally take place from spring, through summer and into the autumn, over a period of several years if the full mycotas of the sites are to be demonstrated.

A broader programme of survey throughout Wales would also be advantageous so that a fuller assessment of the distribution of mycologically important grasslands of the Principality might be made. This would be a valuable contribution to the process of inventory-taking in order to evaluate fungal biodiversity. It would also make a significant contribution to various obligations under the UK Biodiversity Action Group.

The methods of evaluation of conservation value used in this survey have been shown to be workable and informative in this project. A more wide-ranging analysis of the data from the BMS waxcap grassland survey, using these and other similar methods, would produce a clearer, quantitative picture of the important grasslands for fungi in the British Isles, and their distribution. This could be a valuable contribution towards an inventory and comparative study of waxcap grasslands in Europe.

Acknowledgements

This chapter is based on a report published by the Countryside Council for Wales (Rotheroe, 1999) and the author is grateful for their permission to publish this revised version and for their financing of the project. He is also indebted to the National Trust and the Hafod Trust who provided financial support for similar projects. He is grateful for the valuable suggestions on modifications to the text from Vincent Fleming and for the help, advice and support from the following individuals: Jamie Bevan, David Boertmann, Penny David, Shelley Evans, Gareth Griffith, John Bjarne Jordal, Janet Moseley, John Powell, John Savidge, Michael Smith, David Stevens, Nigel Stringer, John Wildig and Ray Woods.

References

Anon. (1999). *UK Biodiversity Group Tranche 2 Action Plans: Volume III – Plants and Fungi*. English Nature: Peterborough.

Arnolds, E. & de Vries, B. (1993). Conservation of fungi in Europe. In *Fungi of Europe: Investigation, Recording and Conservation* (ed. D. N. Pegler, L. Boddy, B. Ing & P. M. Kirk), pp. 211–230. Royal Botanic Gardens: Kew.

Boertmann, D. (1995). *The Genus Hygrocybe. Fungi of Northern Europe 1*. Svampetryk: Greve, Denmark.

Feehan, J. & McHugh, R. (1992). The Curragh of Kildare as a *Hygrocybe* grassland. *Irish Naturalists Journal* **24**(1), 13–17.

Hawksworth, D. L., Kirk, P. M., Sutton, B. C. & Pegler, D. N. (1995). *Ainsworth & Bisby's Dictionary of the Fungi*. Eighth edn, 616 pp. CAB International: Wallingford, Oxon.

Henrici, A. (1996). *Waxcap-grassland Fungi*. British Mycological Society: Stourbridge, UK.

Hodgetts, N. (1992). *Guidelines for Selection of Biological SSSIs: Non-vascular Plants*. Joint Nature Conservation Committee: Peterborough.

Ing, B. (1992). A provisional Red Data List of British Fungi. *Mycologist* **6**, 124–128.

Ing, B. (1993). Towards a red list of endangered European macrofungi. In *Fungi in Europe: Investigations, Recording and Conservation* (ed. D. N. Pegler, L. Boddy, B. Ing & P. M. Kirk), pp. 231–237. Royal Botanic Gardens: Kew.

Keizer, P. J. (1993). The influence of nature management on the macromycete flora. *In Fungi of Europe: Investigation, Recording and Conservation* (ed. D. N. Pegler, L. Boddy, B. Ing & P. M. Kirk), pp. 251–269. Royal Botanic Gardens: Kew.

Noordeloos, M. (1992). *Entoloma s.l.* Libreria editrice Giovanna Biella: Saronno.

Rald, E. (1985). Vokshatte som indikatorarter for mykologisk vaerdifulde overdrevslokaliteter. *Svampe* **11**, 1–9.

Rodwell, J. S. (1991 *et seq.*). *British Plant Communities*. Cambridge University Press: Cambridge, UK.

Rotheroe, M. (1997). *A Mycological Study of NVC Grassland Communities*. JNCC Contract Report No. F76–01–50. Joint Nature Conservation Committee: Peterborough.

Rotheroe, M. (1998). *The Rare Mushrooms of Wales*. Unpublished report: Countryside Council for Wales.

Rotheroe, M. (1999). *Mycological Survey of Selected Semi-Natural Grasslands in Carmarthenshire*. Countryside Council for Wales: Bangor.

Rotheroe, M., Newton, A., Evans, S. & Feehan, J. (1996). Waxcap-grassland survey. *Mycologist* **10**, 23–25.

11

Grasslands in the coastal dunes: the effect of nature management on the mycota

MARIJKE M. NAUTA & LEO M. JALINK

Introduction

Many regions of the dunes along the coast of The Netherlands are now in use as part of the drinking water supply chain and are therefore well protected. The Amsterdam Waterworks Dunes, situated south and west of Haarlem, deliver drinking water to the city of Amsterdam. The site is owned and managed by the municipality of Amsterdam. In this dune field of 34 km^2 the mycota of two, locally decalcified, natural grasslands has been studied since 1986 (Becker & Baeyens, 1992; Nauta & Jalink, 1996). The management consists of summer grazing by cattle in one site (Eiland van Rolvers), and yearly mowing and removal of the hay in the other (Groot Zwarteveld).

History

The area of the Amsterdam Waterworks Dunes consists mainly of young dunes, formed during several periods of blowing sand from the eleventh century onwards. The soils differ in calcium content, depending on the period in which they were formed, and this can often be seen from the vegetation. For instance, the *Hippophae rhamnoides* scrub, developed in the central part of the Amsterdam Waterworks Dunes from around 1500–1600 onwards, has usually developed on more or less calcareous soil. From early days humans have used the dunes. There is evidence that from 1500 onwards, wet dune slacks have been grazed in summer by cattle and sheep.

In the early nineteenth century the dune area also became popular for growing potatoes. Though this was done only for a short period, the fields can still be recognised at several places in the dunes. These fields were fertilised with organic debris and ploughed and are now the most decalcified sites.

In the course of centuries, a sweet-water reservoir has been formed under the dunes, limited in the west by sea water. The area of the Amsterdam Waterworks has served as a water supply for the city of Amsterdam from 1851. Before that time the dunes were so wet that it was possible to skate from Zandvoort to The Hague in winter. After this time a gradual drying has occurred, not only because of the withdrawal of drinking water, but also because of the reclamation of the Haarlemmermeer polder and the removal of sand on the eastern side of the dunes. During World War II many woods were cleared in the dunes, resulting in enhanced drifting of sand.

Because of the noticeable drying of the dunes, infiltration with river water started in 1957, not because of any noble nature maintenance argument, but to secure the water supply for Amsterdam. For this purpose many canals were made and the unpurified water from the river Rhine was led into the dunes in the west, to be purified by the dunes and withdrawn in the east. The unpurified water caused not only a re-establishment of the groundwater water table, but also an enormous increase of certain grasses like *Calamagrostis epigejos* and nutriphilous plants like *Cirsium*. The original wet dune slacks and grasslands in the dunes were soon overgrown with high vegetation of *Calamagrostis epigejos* and *Carex arenaria*. Therefore, the Water Works Company took the decision to purify the infiltrated river water and later began a management programme involving mowing of several grasslands, in the hope this would restore the original vegetation.

Groot Zwarteveld and Eiland van Rolvers

Groot Zwarteveld (in English 'large black field') was in former times a wet, grass-rich dune slack. It was in use by farmers from 1850 for cattle grazing and potato growing. The field owes its name to the thin layers of black peat at the surface, which are no longer visible. In a period with strong winds in the past the valley was blown out and the sand drifted eastward. The soil is locally decalcified.

The potato fields were deserted again around 1900, but the cattle grazing continued until World War II, when the farmhouse was destroyed. Until 1960 the field was described as 'dry and dusty'. Because of the infiltration with river water, the groundwater water table reached its original level of 7 m above the usual reference point which is the elevation of Amsterdam (referred to as NAP) again in 1975, involving the previously mentioned increase of *Calamagrostis*. The area, amounting to 12 ha, has been mown two-yearly since 1975, and the wettest area is mown every year. A high groundwater level also results from a reduced withdrawal of water, and a rain-water lens (a lens-shaped body of rain water floating on the groundwater) is usually present all year round. Rain water is more acidic than groundwater. The infiltrated river water also influences the western part. In the wettest part some *Sphagnum* vegetation occurs.

Eiland van Rolvers (Rolvers' Island) as it is called today, was formerly the eastern part of Groot Zwarteveld. The Rolvers family, who grazed their cattle in the area, inhabited it. It consisted of a wet dune slack and drier dunes with *Hippophae*. From 1975 to 1984 part of the area was mown every second year, and since 1985 it has been grazed in summer by seven yearlings. The area covers 36 ha, and also has a rain-water lens.

Research

The vegetation in both areas has been examined since 1975 and this has shown positive effects of mowing on the composition of the vegetation. As an experiment, in a part of the area the mowing was replaced by grazing in 1985. Grazing was thought to be more natural and less expensive than mowing. In the same period grazing was introduced in several other dune areas as a result of a changing attitude to dune management. At the request of the Amsterdam Waterworks volunteers started to prepare an inventory of all sorts of organisms in 1985 in order to evaluate the effect of both forms of management. For the mycota, the emphasis of the research was laid on the fungi indicative of nutrient-poor grasslands, like waxcaps (*Hygrocybe*), species of *Entoloma*, clavarioid fungi (*Clavaria*, *Clavulinopsis*) and earth-tongues (*Geoglossum*, *Trichoglossum*, *Microglossum*). No permanent plots were installed but the whole grassland area was investigated several times a year. Unfortunately no data on the mycota are available from previous times.

Waxcap grasslands

Grassland fungi are the most striking aspects of the mycota in the two areas. Although the numbers of taxa show large annual fluctuations, there has been a general increase in the number of taxa of *Hygrocybe* in both areas (Fig. 11.1). Within 20 years both areas have developed into good waxcap grasslands. Waxcap grassland is a term invented by Schweers

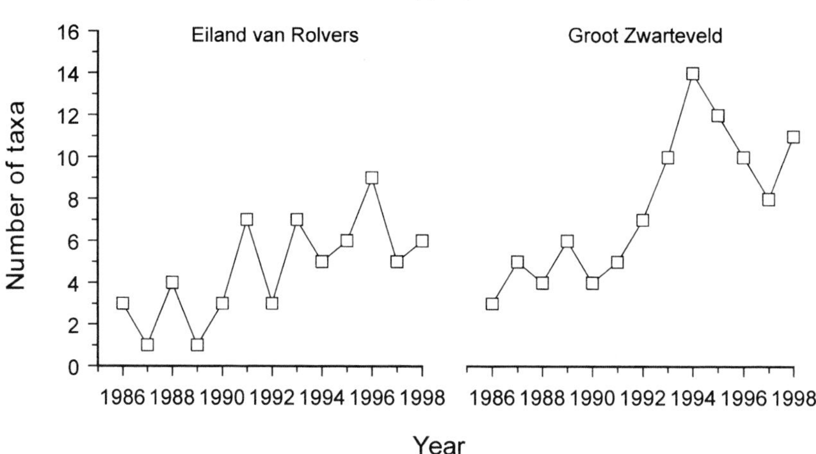

Fig. 11.1. Number of taxa of waxcaps, *Hygrocybe*, found in Eiland van Rolvers and Groot Zwarteveld each year.

(1949) who noticed that in nutrient poor grasslands in The Netherlands a certain mycota was present composed of species of *Hygrocybe*, certain *Entoloma* species and earth-tongues. Grassland fungi belong to one of the most reduced and threatened groups in The Netherlands (Nauta & Vellinga, 1995).

On Eiland van Rolvers a total of 13 taxa of *Hygrocybe* were found. The Groot Zwarteveld, with a total of 17 taxa of *Hygrocybe*, is one of the best waxcap grasslands in The Netherlands. Clavarioid fungi such as *Clavulinopsis luteo-alba*, *C. helveola* and *C. laeticolor* occur as accompanying species. Besides an increase in number of taxa in both sites, a spectacular increase in the number of fruit bodies of *Hygrocybe* and Clavarioid fungi was observed. Unfortunately there is no clear pattern in number of taxa or fruit bodies of grassland *Entoloma* (Fig. 11.2).

Differences

There are also a number of differences between the two investigated areas. More taxa of *Hygrocybe* occur in the mown area (Groot Zwarteveld), and they are in general slightly more acidophilic (Table 11.1). In the mown area a sudden increase is shown in 1993 in the species *Hygrocybe acutoconica*, *H. helobia*, *H. insipida*, *H. laeta*, *H. psittacina*. In the grazed area (Eiland van Rolvers) the increase in number of taxa has taken place

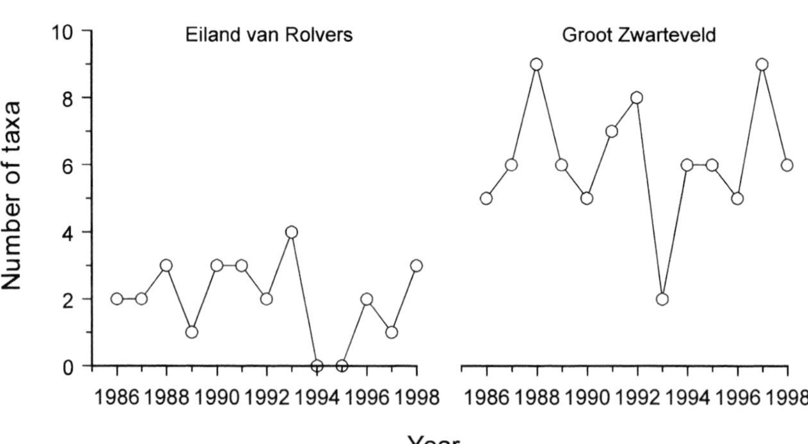

Fig. 11.2. Number of taxa of grassland *Entoloma* found in Eiland van Rolvers and Groot Zwarteveld each year.

more gradually, but in 1991 *H. conica* and *insipida* were suddenly found.

In the grazed area more taxa and especially far more fruit bodies of earth-tongues were found. The same applies to the clavarioid fungi.

An explanation of the differences can be found in the difference in landscape, management and vegetation. Eiland van Rolvers shows a more varied vegetation on more broken ground, also with dune-scrub, *Crataegus* and *Hippophae*. In particular, the small hilly area is drier than Groot Zwarteveld. The richness in clavarioid fungi is connected with the drier parts of the terrain; the richness in earth-tongues is caused by the presence of gradients in the soil below the hills and at the margin of scrub. Groot Zwarteveld is more decalcified because of the former use as potato fields, which is manifested for example in the occurrence of *Sphagnum* in the small pools, and the area is slightly wetter. This can explain the difference in species of *Hygrocybe*.

Cattle do not graze rushes and the moist parts of Eiland van Rolvers are partly overgrown by rushes. Protected by dense rush vegetation, seedlings of birches have grown up, resulting in the development of birchwood on Eiland van Rolvers and an increase of important and threatened ectomycorrhizal fungi, like *Lactarius vietus*.

Grazing also has a direct enriching effect on the mycota, namely fungi on dung. In the grazed area many coprophilous fungi were found which have been under threat in The Netherlands. For example, *Panaeolus*

Table 11.1. *Number of years in which taxa of* Hygrocybe *occurred in Groot Zwarteveld and Eiland van Rolvers*

Taxa of *Hygrocybe*	Groot Zwarteveld	Eiland van Rolvers
acutoconica	5	
cantharellus	1	
ceracea	12	8
coccinea	2	3
conica var. *conica*	13	6
conica var. *conicopalustris*	2	
conica var. *chloroides*	2	
glutinipes	4	
helobia	6	2
insipida	6	6
laeta	5	5
luteolaeta	8	
miniata var. *miniata*	13	13
miniata var. *mollis*	3	1
pratensis var. *pallida*		1
psittacina	5	7
russocoriaceus	1	1
virgineus var. *virgineus*	10	6
virgineus var. *fuscescens*		1

fimiputris is a species that seriously declined during the sixties and seventies (Nauta & Vellinga, 1995), but has recently begun to recover. It is a species that occurs on fibre-rich cattle dung, a substrate that had become very scarce in The Netherlands in the past few decades. Thanks to an increase in cattle grazed on nutrient-poor grasslands in the dunes, and nature reserves more inland, several formerly threatened species are showing an increase. Other interesting finds on the dung are *Panaeolus speciosus* (only recently discovered in The Netherlands; Van de Bergh & Noordeloos, 1996), and the rare *Peziza bovina*.

Mowing or grazing?

From the mycological point of view, both mowing and grazing have proved to be very valuable management methods. Both result in a rich mycota, although the species spectrum is different. As a result of our studies, the management of the Municipal Waterworks of Amsterdam has been advised to continue both methods. In general, mowing gives a larger area of grassland and, consequently, more habitat for grassland fungi.

Grazing results in a mosaic of vegetation types ranging from grassland, scrub and small woodlands. This gives opportunities to a wider range of fungi, including ectomycorrhizal fungi of woods and special groups like dung fungi. As far as cost is concerned, grazing is generally cheaper than mowing, especially since the hay is relatively useless in terms of nutrition for cattle, but grazing does not prevent development of scrub and forest. However, if grazing is stopped now, the area will probably become rapidly overgrown again.

On Groot Zwarteveld a stable situation has now developed, with a short, grassy dune slack vegetation, whilst on Eiland van Rolvers the vegetation is still under development. Because of the expansion of rushes and development of the *Betula* wood, some of the grassland in the western part is becoming overgrown. In the central and eastern part the vegetation is far more varied, with nicely developed scrub of *Salix repens* and *Hippophae* on the more calcareous parts.

Future

A difficult question in nature management concerns the future. Nowadays there is a strong tendency to let nature rule itself. In the case of grasslands in the dunes there have been some recent developments caused by human intervention which have not been beneficial for biodiversity, for instance. The large fields of *Calamagrostis* and *Carex arenaria* were caused by humans, and bringing it back to a situation that existed around 1500 does not seem unnatural. Of course, for many groups of (lower) fungi and other organisms it is not known what the effect will be. But in general it is thought that biodiversity will benefit. If natural processes are allowed to do their job in the dunes the climax vegetation is the formation of dune scrub and woods. The result of this would probably also be a biota rich in fungal species.

References

Becker, A. & Baeyens, G. (1992). Wasplaten en andere paddestoelen in een vochtige vallei van de Amsterdamse Waterleidingduinen. *De Levende Natuur* **93**, 111–117.

Nauta, M. M. & Jalink, L. M. (1996). *Ontwikkelingen in de mycoflora onder invloed van begrazing op het Eiland van Rolvers.* Gemeentewaterleidingen Amsterdam: Amsterdam.

Nauta, M. M. & Vellinga, E. C. (1995). *Atlas van Nederlandse paddestoelen.* A. A. Balkema Uitgevers B. V.: Rotterdam.

Schweers, A. C. S. (1949). De Hygrophorusweide, een associatie. *Fungus* **19** (2), 17–18.

Van de Bergh, F. & Noordeloos, M. E. (1996). *Panaeolus speciosus* P. D. Orton: nieuw voor Nederland. *Coolia* **39**, 74–77.

12

The conservation of fungi on reserves managed by the Royal Society for the Protection of Birds (RSPB)

MARTIN ALLISON

Introduction

The Royal Society for the Protection of Birds (RSPB) is the largest wildlife conservation charity in Europe with over one million members. It manages over 150 nature reserves throughout the UK, covering more than 108 000 hectares. Away from the reserves the Society safeguards sites and species of conservation importance through research, lobbying and education, whilst outside the UK the RSPB is working with Birdlife International Partners on major conservation projects in 18 countries in Europe, Africa and Asia. The RSPB, together with voluntary organisations, has produced a practical guide to conserving the UK's biodiversity that has been adopted at Government level.

The Society is committed to the management of nonavian taxa on its reserves, and is UK champion for nonbird species such as the medicinal leech, *Hirudo medicinalis*. The Society's interest in other biotic groups has been somewhat biased towards invertebrates and lower plants, for the most part to the exclusion of the fungi. This latter fact is understandable given the difficulties with identification and the complex taxonomy within mycology. In recent years an interest in the fungi on RSPB reserves has been encouraged by one or two active field mycologists, and has gradually led to the development of an in-house strategy for their conservation.

Threatened in Europe

A rapid decline of fungi across northern Europe has been reported since the 1980s. Before discussing habitats and species on RSPB reserves the European situation needs to be understood.

An estimated 5106 species of macromycetes are thought to occur in northern Europe. Of these, 2658 (52%) are found on one or more European Red Data list (Arnolds & de Vries, 1993). Proportionately more extinct or endangered species on Red Data lists occur in highly industrialised and/or densely populated areas such as The Netherlands, Germany and Czechoslovakia. Nordic countries and Britain have a larger proportion of vulnerable, rare or potentially threatened fungi.

No less than 55% of the European agaric fungi appear on one or more Red Data lists (Arnolds & de Vries, 1993). Many of these species are ectomycorrhizal and their ecological role is considered vital to the health of the forests (Rayner, 1993). Some forests in industrialised areas of Europe are showing sharp declines in their ectomycorrhizal species; this often occurs before any signs of deterioration are detected in the host tree species. Such declines particularly occur within hillside forests on poor soils, an ecosystem commonly found in north Europe, where the habitat is affected by acidification from air pollution (Fellner, 1993). This causes decreases both in numbers of species of fungi and numbers of fruiting bodies.

Of the fungal species characteristic of dry nutrient-poor grassland, 78% are threatened in The Netherlands (Arnolds, 1989). These grassland species were once widespread, but changing agricultural methods coupled with excessive use of fertilisers has caused a worrying decline in north Europe generally. Mycologically rich grassland is often several centuries old but it can be mirrored on sensitively managed old lawns.

The main external threats to European fungal communities can thus be summarised as air pollution, change of land use and agricultural methods, and urban encroachment. Commercial picking and overzealous clearing up of dead wood may also seriously threaten fungi and organisms such as fungus gnats, beetles and slugs that depend upon them. All of these threats are relevant in the British Isles and on RSPB nature reserves.

Recording of fungi at RSPB reserves

Macrofungi fall into three broad habitat categories. The ectomycorrhizal species associating with tree roots; the saprotrophic species of fungi living on dead organic matter and recycling essential nutrients, and the lignicolous species inhabiting wood and thus creating the hollow trees needed for hole nesting birds. For the purposes of describing fungi at RSPB reserves, the species will be restricted to the agarics and boletes, the Aphyllophorales and selected ascomycetes.

Many records on RSPB reserves originate from organised forays where

Table 12.1. *Important RSPB reserves for fungi showing totals of Red Data list species*

Reserve	Score[a]	R	V	E	A	B	C	D
Abernethy	180	4	38	7	2	8	21	2
Tudeley Woods	109	4	15	2	1	5	25	
Lake Vyrnwy	28	1	6			1	6	
Geltsdale	24		5			1	5	1
Blean Woods	19		1	1		1	5	1
Northward Hill	18		1	1		1	5	
Killiekrankie	17	1	3				5	
Stour Wood	16		5				3	
The Lodge	15		1				6	
Minsmere	14		1			2	3	
Leighton Moss	13	1					6	
Wolves Wood	12			1		1	3	
Loch of Strathbeg	11		1			1	3	

British Red Data (preliminary) list Categories: R, Rare; V, Vulnerable; E, Endangered.
European Red Data list Categories: A, widespread losses, rapidly declining populations, many national extinctions, high-level concern; B, widespread losses, evidence of steady decline, some national extinctions, medium-level concern; C, widespread but scattered populations, fewer extinctions, lower-level concern; D, local losses, some extinctions but mainly at edge of geographical range.
[a]Score weightings based on numerical values attributed to differing rarity categories.

voucher specimens of the rarer species have been deposited in herbaria. Other lists are supplied by regional amateur mycologists, or by the reserve staff themselves. Individual mycologists tend to concentrate on certain taxa that interest them. This is somewhat reflected in the choice of Red Data list species.

Forty-four RSPB reserves have contributed lists totalling over 1860 species. Of these, 81 species occur on the European, and 77 on the British provisional Red Data lists (Ing, 1992). Important reserves for threatened fungi are shown in Table 12.1.

Some reserves have been studied in much greater detail than other potentially rich reserves: Abernethy Forest, Highland; Tudeley Woods, Kent; Leighton Moss, Cumbria; Nagshead, Gloucestershire, and The Lodge, Bedfordshire, are among those that have been better studied (see Fig. 12.1). Potentially rich sites include: Ynys hir, Powys; Wye-Elan Valley, Powys; Geltsdale, Cumbria; Coombes Valley, Staffordshire; Hobbister, Orkney.

Of the habitats represented on reserves, the most important mycologi-

Fig. 12.1. Locations of RSPB reserves that are important for fungal conservation.

cally are Caledonian pine forest; the nutrient-poor, unimproved grasslands of mainly northern and western Britain; and the semi-natural ancient woodlands of southeast England. Sand dunes and western oak woodlands also offer potential but, unfortunately, few data were available from reserves.

Geographical distribution at reserves

Generally, fungi are widespread across the whole of Britain, and indeed many British species are transglobal through temperate regions. Some species, however, show distinct geographical limitations. Of the more interesting species at RSPB reserves, horse hoof fungus *Fomes fomentarius*, *Rozites caperatus* (commonly found in Highland coniferous forests), and the endangered *Lactarius musteus* found at Abernethy Forest, are all predominantly fungi of northern Britain. By contrast, *Russula aurata*, *R. pseudointegra*, *Boletus pseudoregius*, *B. aereus* and *Leccinum crocipodium* are among the fungi that show a preference for warmer southern woodlands. There are a few predominantly western species, the most distinctive of which is the delicate grey-blue fairy bonnet *Mycena pseudocorticola*, recorded from Wye-Elan Valley, Powys, and West Sedgemoor, Somerset.

Important habitats for fungi at RSPB reserves

Caledonian pine forest

The Abernethy Forest reserve (Fig. 12.1) boasts one of the largest remaining areas of Caledonian pine forest. It is an important habitat for many rare and threatened species and assemblages of fungi, both within the forest and on the surrounding moorland. The reserve supports 33 European and 49 British Red Data list species; seven of the latter are considered endangered.

The milk cap *Lactarius musteus* and the agaric *Tricholoma focale* are two endangered species on the British list associated with Scots pine *Pinus sylvestris*. No fewer than 10 stipitate hydnoid fungi can be found at Abernethy; all are on the Red Data lists and two, *Bankera fuligineoalba* and *Phellodon confluens*, are endangered in both the UK and Europe (Table 12.2). Another endangered species is the purple cup-fungus *Ombrophila violacea* that grows on decaying leaves in boggy places.

Abernethy Forest also supports impressive assemblages of some important genera, notably 55 species of *Cortinarius* including 9 on the British Red Data list, 40 species of *Russula* (7 on the Red Data list) and 14 species of *Tricholoma* (3 Red Data list species).

Semi-natural ancient woodland

These predominantly southern woodlands are the strongholds of the ectomycorrhizal species in Britain and harbour important and increasingly rare assemblages of fungi. The most representative reserve for this group is at Tudeley Woods in Kent (Fig. 12.1), which supports 26 European and 16 British Red Data list species; two of the latter are considered endangered.

Table 12.2. *Stipitate hydnoid species on RSPB reserves*

Species	Red Data list categories	Reserve
Bankera fulgineoalba	A, E	Abernethy
Hydnellum aurantiacum	B, V	Abernethy
H. caeruleum	B, V	Abernethy
H. concrescens	C, V	Abernethy, Tudeley Woods
H. peckii	B, E	Abernethy
H. scrobiculatum	B, E	Blean Woods, Tudeley Woods
H. spongiosipes	C, R	Tudeley Woods
Hydnum repandum	widespread	
Phellodon confluens	A, E	Abernethy
P. melaleuceum	C, V	Abernethy, Northward Hill
P. niger	B, R	Tudeley Woods
P. tomentosum	B, E	Abernethy, Northward Hill
Sarcodon glaucopus	B	Abernethy
S. imbricatum	C, V	Abernethy, Tudeley Woods
S. scabrosum	B, E	Tudeley Woods

For Red Data list categories see Table 12.1.

Boletus pseudoregius is a rare and possibly endangered European fungus associated with warm, southern oak woodland and was first noted in 1995. Along with *B. regius* and *B. speciosus*, this taxonomically difficult group is currently under review (Marren, 1998). Other European Red Data list boletes at Tudeley are *B. aereus*, *B. impolitus*, *B. queletii*, *B. radicans* and *B. appendiculatus*. These fungi indicate a continuity of woodland cover over many centuries. The second endangered British species at Tudeley Woods is the hydnoid *Sarcodon scabrosus*. Hedgehog fungi associated with RSPB woodlands in Kent are also found at Blean Wood and Northward Hill (Fig. 12.1).

Other rare European species of southern woodlands are *Cortinarius bulliardii* from Fore Wood, East Sussex (Fig. 12.1), and the attractive *C. violaceus* at the small Barfold Copse reserve in Surrey. Minsmere in Suffolk holds *Boletus queletii* and *Hericium erinaceus*, whilst *Boletus impolitus* is found in Blean Woods. The black-capped bolete *Strobilomyces floccopus* is rare outside the Severn and Thames basins but occurs at a few outlying stations including Tudeley Woods and Fore Wood, East Sussex. *Entoloma euchroum* is a small blue fungus that is included on eight of the eleven individual country Red Data lists across Europe, but not on the British one. It occurs at Stour Wood and Wolves Wood in Essex as well as at Tudeley where it grows only at the base of coppiced hazel *Corylus avellana*.

Unimproved nutrient-poor grassland

This habitat is mainly represented in the north and west of Britain, where it has suffered less from intensive agriculture than in lowland Britain. Such grassland is represented in the upland pastures of: Abernethy; Killiecrankie, Perthshire; Geltsdale, Cumbria; Lake Vyrnwy, Powys (Fig. 12.1). Important groups of fungi associated with these pastures are the waxcaps and the entolomas. Many of these species are declining at alarming rates in some parts of Europe.

Of the 43 waxcap species recorded from RSPB reserves, 28 are found on unimproved pasture at Abernethy, 20 at Lake Vyrnwy, 16 at Geltsdale, 15 each at Killiecrankie and Leighton Moss, Lancashire. Most of the 13 waxcaps that occur at Tudeley Woods are found within the coppiced woodland, including the Red Data listed *Hygrocybe calyptriformis*. It is assumed that the presence of these waxcap species reflects the fact that the woodland was once grazed as wood pasture when extensive grassy swards would have been present. Forty-nine species of *Entoloma* have been recorded at RSPB reserves: 25 occur at Lake Vyrnwy, and 24 at Abernethy. Other declining grassland fungi are the clavarioid fungi, 16 of which occur at RSPB reserves. Of the four grassland earth-tongues recorded at reserves, three are European Red Data species, *Microglossum viride* at Killiecrankie, *Geoglossum nigritum* at Leighton Moss and *Trichoglossum hirsutum* at Leighton Moss, Tudeley Woods and on the lawns at The Lodge, Bedfordshire, which is headquarters to the RSPB. Such well-established lawns can be rich in fungi such as *Hygrocybe*, *Entoloma*, *Clavaria* and *Geoglossum*.

Heathlands

Heathlands support a few uncommon species of fungi, but are mainly noted for their rare bird, higher plant and reptile assemblages. Of the more interesting ones on reserves are *Clavaria argillacea* (Abernethy, The Lodge, Minsmere, Vane Farm and Loch of Strathbeg), *Cantharellula umbonata* (Vane Farm and Lake Vyrnwy) and *Mycena uracea* (Stour Wood, Blean Woods, Northward Hill, The Lodge and Abernethy). This latter species is often associated with burnt ground and is found particularly around heather shoots.

Plantations

Habitats within nonindigenous woodlands should not be ignored. Of particular worth are conifer plantations and sweet chestnut coppice *Casta-*

Table 12.3. *Regional importance of selected taxa from two RSPB reserves in southeast England shown as a percentage of the total species for southeast England*

Taxon	Tudeley (%)	Blean (%)	Total for the two reserves (%)
Lactarius	68	44	72
Russula	53	35	55
Mycena	43	31	49
Pluteus	32	14	41

nea sativa, both occasionally supporting rare species and communities of fungi. Sweet chestnut coppice in southern England often harbours species found in native mixed oakwoods, particularly the boletes, russulas and red data hydnoid species. The rare *Strobilomyces floccopus* mentioned earlier has been found in chestnut coppice. A resupinate corticioid species, *Byssocorticium efibulatum*, was recorded from the underside of chestnut logs at Tudeley Woods in 1994 and was a new British record (Roberts, 1994).

Introduced conifers play host to many species of important genera such as *Lepiota* and *Cortinarius*. At The Lodge the European Red Data list species *Volvariella caesiotincta*, *Geastrum pectinatum*, *G. triplex* and *Hygrophorus hypothejus* are found alongside several scarce lepiotas within the softwood plantations. Accompanying these is the very rare *Hygrophorus speciosus*, being newly recorded for Britain in 1984, and the only known site in the country for this fungus.

Regional importance of two RSPB reserves in southeast England for fungi

Comparisons with major groups of agarics in southeast England are possible with the publication of an updated checklist for East and West Sussex, Kent and Surrey (Dennis, 1995). For the two large RSPB woodland complexes of Tudeley Woods and Blean Woods, both in Kent, regional and national percentages of selected groups of fungi are shown in Table 12.3. The two reserves together support 90% of the *Lactarius* species and 72% of *Russula* that occur in southeast England. Tudeley Woods with its extensive derelict coppice and much fallen mature timber, including elm and beech, is a favoured site for lignicolous *Pluteus* fungi.

A further comparison with a major mycological site, Bedgbury Pinetum in Kent, shows that Tudeley Woods supports 26% of the southeast

England agaric biota, compared with 32% at Bedgbury (Weightman, 1993). Tudeley also supports 26% of the region's aphyllophoralean fungi.

The conservation of fungi

The conservation of many fungi at nature reserves tends to be limited by the fact that most threats are beyond the control of the site's management. By far the greatest of these is acidification and eutrophication from airborne pollutants, although Britain suffers less than most mainland European countries. This can best be tackled through international environmental policy. Some experimental work is in progress in other parts of Europe where soil manipulation has been monitored and compared with fungal productivity (Baar & Kuyper, 1993). Results from this work are not yet conclusive but do suggest that a rich ectomycorrhizal biota correlates with a thin or absent organic layer. This can be achieved artificially by the removal of litter/mulch layers from the soil, thus helping reverse the trend in fungal decline by removing damaging nitrogenous compounds. However, this type of management could prove more harmful than acidification to ancient woodland soil ecosystems and their saprotrophic fungal species.

To date there are no comparisons available between managed and nonintervention forestry in this country with regards to fungal communities (Keizer, 1993). Opportunities present themselves on extensive woodland reserves such as Tudeley, Blean or Abernethy where nonintervention and actively managed compartments coexist. In order to make any study feasible either large permanent quadrats or substantial transects need to be set up and monitored for a minimum ten year period. This would ideally merge with studies of other biotic groups such as National Vegetation Classification (NVC) vegetation recording or Institute of Terrestrial Ecology (ITE) butterfly and other invertebrate monitoring programmes.

Wood-inhabiting fungi, along with many invertebrates, are threatened by timber management/tidying up operations. Apart from essential safety measures, dead wood is left *in situ*, standing or fallen, as important habitat (Green & Alexander, 1993). Although few RSPB reserves hold areas of old parkland, which are often internationally important for fungi, there are many overmature individual trees within wooded areas that may well hold important lignicolous species. Some uncommon bracket fungi rely on access to the heartwood of trees to survive. A proportion of healthy mature trees is always retained in perpetuity on reserves as potential future hosts for these fungi. Heartwood species do not kill off the host trees as the outer sapwood is unharmed and continues to feed the tree, thus producing the hollow oaks and beeches of sites such as Windsor Great Park.

Management of the remaining unimproved nutrient-low grassland on RSPB reserves is similar to managing for their bird or botanical interest. A regime of low-density grazing, or mowing/forage harvesting followed by grazing, is generally recommended. Fertilisers are avoided other than natural droppings from livestock.

The management at Abernethy reserve is especially delicate as it is a key site for both Caledonian pine forest species and fungi of unimproved pasture. A balance needs to be sought here to stop overencroachment onto the important *Hygrocybe*/*Entoloma* grassland communities, whilst ensuring the pinewoods are given full protection and allowed to perpetuate for the future.

Large-scale collection of fungi can be indirectly damaging. The penny bun bolete *Boletus edulis* is considered threatened through overpicking in Poland. Research has shown, however, that collection may actually stimulate the growth of fruit bodies. The problems arise more from soil compaction through trampling, and the removal of habitat for the mammals and invertebrates that feed on fungi. In Britain, commercial and private collecting of the penny bun and other species is increasing through exposure via natural history and cookery programmes on television.

Internal threats to fungi on reserves

Potentially more threatening than human trampling is the modern practice of using heavy timber extraction machinery on reserves, which can lead to extensive ground disturbance and soil compaction. This can sever the mycorrhizal links between tree and fungus, causing long-term ill health within the forest ecosystem.

Conflicts may arise within conservation management when one group of animals or plants is favoured above another. An example encountered on RSPB reserves is short-term coppice and ride management to encourage butterflies and other invertebrates. This creates large areas of dense scrub and bramble which limits the fruiting of sometimes rare species of fungi which once thrived in the damp and shaded uncut coppice. This particular situation is normally rectified by allowing the coppice to regrow whilst creating scalloping and glades away from the threatened fungus species.

At the RSPB headquarters in Bedfordshire the rare *Hygrophorus speciosus* was seen to be threatened by a heathland restoration project where trees were to be removed close by the fungus. An island of conifer was therefore retained and later saplings were replanted on the periphery to act as a windbreak against desiccation. Furthermore, an attempt will be made to translocate the fungus into other areas of the reserve through encouraging mycorrhizal links with young planted conifers that will then be removed and transplanted elsewhere.

The RSPB has begun to address these problems through its nonavian biodiversity monitoring programme (Cadbury & Evans, 1995). Species currently within the programme include *Boletus pseudoregius*, *Hygrophorus speciosus*, *Hygrocybe calyptriformis*, *Hericium erinaceum* and several hydnoid species.

Conclusions

In summary, the RSPB hopes to manage for fungi under each reserve's broad policy statements. To that effect, all RSPB reserve management plans will eventually contain reference to British and European Red Data listed species and threatened communities, and include prescriptions for their conservation alongside those of other taxa.

It is difficult to provide precise management prescriptions for such understudied organisms as the macrofungi. As has already been mentioned, many threats are external and habitats have already been destroyed. The rule of thumb must be that, if the fungus is present on the reserve then the habitat is fundamentally suitable and the status quo should be retained. Research should then be encouraged to ensure the populations are stable or increasing and this should include the effects of grazing, the effects of coppicing, the study of fungal communities in nonintervention zones, etc., depending on the habitat. Guidelines should be sought from scientists specialising in field mycology on ways to approach such research.

The important factor in fungal conservation on reserves is that the land managers in charge of each site are aware of the presence and status of fungi and that they retain the long-term traditional management or 'non-management' that has allowed them to survive up to the present. To do this successfully, regular recording is necessary, preferably by knowledgeable amateurs from outside the Society.

The RSPB has taken an essential first step in acknowledging the importance of the fungi on its nature reserves by ensuring the management planning system is aware of their presence and current status, and by including the rarest species and communities in its nonavian biodiversity monitoring programme.

Acknowledgement

Adapted by the author from his contribution to the RSPB Conservation Review 1997 (Allison, 1997).

References

Allison. M. (1997). The importance of RSPB reserves for fungi. *RSPB Conservation Review* **11**, 88–95.

Arnolds, E. (1989). A preliminary red data list of macrofungi in The Netherlands. *Persoonia* **14**, 77–125.

Arnolds, E. & de Vries, B. (1993). Conservation of fungi in Europe. In *Fungi of Europe: Investigation, Recording and Conservation* (ed. D. N. Pegler, L. Boddy, B. Ing & P. H. Kirk), pp. 211–230. Royal Botanic Gardens: Kew.

Baar, J. & Kuyper, T. W. (1993). Litter removal in forests and effects on mycorrhizal fungi. In *Fungi of Europe: Investigation, Recording and Conservation* (ed. D. N. Pegler, L. Boddy, B. Ing & P. H. Kirk), pp. 275–286. Royal Botanic Gardens: Kew.

Cadbury, J. & Evans, C. (1995). The monitoring of non-avian species of conservation importance at RSPB reserves: a strategy. Unpublished report, RSPB: Sandy.

Dennis, R. W. G. (1995). *Fungi of South East England.* Royal Botanic Gardens: Kew.

Fellner, R. (1993). Air pollution and mycorrhizal fungi in Central Europe. In *Fungi of Europe: Investigation, Recording and Conservation* (ed. D. N. Pegler, L. Boddy, B. Ing & P. H. Kirk), pp. 239–250. Royal Botanic Gardens: Kew.

Green, T. & Alexander, K. (1993). Dead wood – eyesore or ecosystem? *Enact* **1**, 11–14.

Ing, B. (1992). A provisional Red Data list of British fungi. *Mycologist* **6**, 124–128.

Ing, B. (1993). Towards a red list of endangered European macrofungi. In *Fungi of Europe: Investigation, Recording and Conservation* (ed. D. N. Pegler, L. Boddy, B. Ing & P. H. Kirk), pp. 231–237. Royal Botanic Gardens: Kew.

Keizer, P. J. (1993). The influence of nature management on the macromycete flora. In *Fungi of Europe: Investigation, Recording and Conservation* (ed. D. N. Pegler, L. Boddy, B. Ing & P. H. Kirk), pp. 251–269. Royal Botanic Gardens: Kew.

Marren, P. (1998). *Boletus regius* Kromb., the Royal Bolete and the pretender *Boletus pseudoregius* Estades. Species Recovery Report no. 111. English Nature/Plantlife.

Rayner, A. D. M. (1993). The fundamental importance of fungi in woodlands. *British Wildlife* **4**, 205–215.

Roberts, P. (1994). Interesting and unusual corticioid fungi from Slapton, Devon II. *Mycologist* **8**, 115–118.

Weightman, J. (1993). The fungi of Bedgbury forest. *Transactions of the Kent Field Club* **13**, 1–28.

13

Strategies for conservation of fungi in the Madonie Park, North Sicily

GIUSEPPE VENTURELLA & SALVATORE LA ROCCA

Introduction

The Madonie Park, covering 40 000 hectares, is one of the most interesting and floristically differentiated areas in the Mediterranean basin. Its territory shows a very high degree of diversity as instanced by about 1600 taxa of the vascular flora with a high percentage of endemic species. The forest vegetation, in the Mediterranean to subatlantic belt, is varied and mainly characterised by evergreen oak (both *Quercus ilex* and *Q. suber*) and deciduous woods with *Quercus pubescens*, *Q. virgiliana*, *Castanea sativa*, *Q. petraea* and *Fagus sylvatica*. Exotic trees are widespread in all the Madonie mountains, including *Pinus halepensis*, *P. pinea*, *P. nigra*, *Cedrus atlantica*, *C. deodara*, *Cupressus sempervirens*, *C. arizonica*, *C. macrocarpa*, *Abies alba*, *A. cephalonica*, *Robinia pseudoacacia* and *Eucalyptus camaldulensis*. The landscape is also composed of rocky environments, grasslands and wetland sites. All these environments are also very rich in cryptogams (lower plants) but knowledge about them is still inadequate.

As regards fungi, it is noteworthy that since the second half of the seventeenth century the local population has exploited the understorey products as a source of food or for income. In the last decade mushroom picking, previously limited to relatively few people, has become a widespread pastime with potentially adverse consequences to mycelium growth and fruit body appearance. This threat, together with the lack of knowledge of Sicilian fungi, causes serious problems for any attempt to safeguard the integrity of the ecosystem. In addition, the concern for fungi by local authorities and the scientific community is still very slight.

As a first step towards conservation of the Madonie mycota, a census project was started in 1991 (Venturella, 1992). The aims of this programme are to increase the knowledge on the distribution and ecology of the

Sicilian macrofungi, to locate rare and endangered taxa and to propose conservative actions.

Data recorded

At present 614 taxa included in 200 genera belonging to 79 families have been recorded. This corresponds to 50% of the known Sicilian mycota. Considering that in a previous paper (Venturella & Mazzola, 1991) concerning the state of the mycological exploration in Sicily, only 65 taxa belonging to 49 genera were reported for the Madonie Mountains, it is clear that the new data represent a major improvement in knowledge. Among the taxa recorded three main groups were identified: taxa with wide distribution and ecology, taxa strictly linked to some specific environments, and taxa which are rare and/or distributed only in restricted areas.

Among the taxa showing a wide distribution and ecology, *Amanita pantherina*, *A. phalloides* and *Lepista nuda* are widespread. They occur, mainly during autumn and winter, in all the woodlands and are abundantly represented in terms of their numbers of fruit bodies. *Mycena pura* and *Schizophyllum commune* have similarly wide distribution as these taxa but their fruiting period was extended all year long. The fruiting period of *Russula delica*, *R. chloroides* and *R. foetens* is, however, restricted to autumn. These taxa also show a wide distribution, being frequently recorded.

Agaricus silvicola, *Clitocybe nebularis*, *C. odora*, *Entoloma lividum* and *Lactarius chrysorrheus* are widely distributed in the broad-leaved woods and in many other localities within the area investigated. *Amanita rubescens*, *Inocybe rimosa* and *Tricholoma scalpturatum* are also widely distributed, but their appearance is limited as far as the number of records is concerned. Among the boletes, *Boletus aereus* is the most widely distributed and recorded species. In marked contrast with its occurrence in the Apennine woods, it is recorded not only in evergreen oak woods and thermophilic broad-leaved woods but also in the beech woods which are distributed in the northern and eastern parts of the island from 1200 to 2000 m.

Of taxa strictly linked to particular environments, *Hebeloma cistophilum*, *Lactarius tesquorum* and *Leccinum corsicum* are widespread in the areas characterised by the presence of *Cistus creticus*, *C. monspeliensis* and *C. salvifolius*. *Cortinarius cedretorum*, *Geopora sumneriana* and *Tricholoma tridentinum* var. *cedretorum* (Venturella, 1995) are prevalent in the woods with *Cedrus atlantica* and *C. deodara*, which were introduced throughout the area investigated since the 1960s, after the reforestation carried out by the local Forestry Administration.

The use of wood for charcoal caused the occurrence of fungi specifically associated with burnt areas such as *Faerberia carbonaria*, *Pholiota highlandensis*, *Hebeloma anthracophilum* and *Tephrocybe atrata*. In areas reforested with *Eucalyptus camaldulensis*, the occurrence of *Setchelliogaster tenuipes* is noteworthy (La Rocca & Anastase, 1997*a*).

Although present knowledge is incomplete, we can go some way to evaluating the state of decline of Sicilian macrofungi by making a comparison between this survey and the provisional European Red Data list of endangered macrofungi published by Ing (1993). As shown in Table 13.1, *Boletus rhodoxanthus*, belonging to Group A, seems not to be declining since it was recorded 23 times during the period of investigation. Of taxa included in Group B, *Boletus impolitus*, *B. queletii* and *Hygrophorus russula* are quite common in the Madonie area, whilst *Entoloma bloxamii*, *Myriostoma coliforme* (La Rocca & Anastase, 1997*b*) and *Boletus fechtneri* also confirm their status as being infrequent in the Mediterranean area.

In Group C, *Boletus radicans* and *B. aereus* are really common, while *Omphalotus olearius* and *Leucopaxillus gentianeus* were frequently recorded in the broad-leaved woods. Of the taxa included in Group D, *Amanita caesarea* was prevalent only in the siliceous areas of the Madonie Mountains, where fruit bodies could be abundant. Among the taxa recorded, *Agaricus bernardii*, *A. pseudopratensis*, *Choiromyces meandriformis*, *Collybia cookei*, *Coprinus vosoustii*, *Entoloma undatum*, *Gyroporus cyanescens*, *Helvella solitaria*, *Lactarius decipiens*, *Melanoleuca pseudovenosa* and *Tricholoma bresadolanum* were infrequent, while *Entoloma atrocoeruleum*, *Macrotyphula fistulosa*, *Ossicaulis lignatilis* and *Pleurotus nebrodensis* were rare not only in the Madonie Mountains but also in all the Sicilian territory.

These taxa are not included in Ing's list (Ing, 1993) but they could be regarded as worthy of protection, at least in Sicily.

Conservation strategies

In parallel with the monitoring and inventory of fungal diversity in the Madonie area, a protection plan for the mycota was proposed (Venturella, 1992). It provides for introduction of rules concerning picking and the trade in fungi within the Madonie Park. In particular, it includes a prohibition of the use of rakes in the searching process, because they can cause damage to the mycelium and roots, and also the use of plastic bags for

Table 13.1 *Taxa included in Ing's list*[a] *compared with the number of records of the Madonie area*

Taxa		Records
Group A	*Boletus regius*	9
	Boletus rhodoxanthus	23
	Boletus satanas	12
Group B	*Aureoboletus gentilis*	3
	Boletus fechtneri	1
	Boletus impolitus	28
	Boletus queletii	26
	Cortinarius bulliardii	3
	Cortinarius violaceus	4
	Entoloma bloxamii	1
	Hericium erinaceus	5
	Hygrophorus russula	18
	Lactarius mairei	7
	Myriostoma coliforme	1
	Phellodon niger	3
	Tricholoma acerbum	12
Group C	*Amanita franchetii*	4
	Astraeus hygrometricus	8
	Boletus aereus	136
	Boletus appendiculatus	4
	Boletus fragrans	11
	Boletus radicans	98
	Cantharellus cinereus	3
	Coriolopsis gallica	2
	Cortinarius cinnabarinus	1
	Cortinarius praestans	9
	Ganoderma resinaceum	4
	Geastrum rufescens	1
	Grifola frondosa	2
	Hygrocybe punicea	3
	Hygrophorus arbustivus	7
	Hygrophorus camarophyllus	1
	Lactarius violascens	1
	Lepiota cortinarius	8
	Leucopaxillus gentianeus	27
	Mutinus caninus	1
	Omphalotus olearius	30
	Oudemansiella pudens	10
	Pisolithus arhizus	8
	Pulcherricium caeruleum	8
	Ramaria botrytis	8
	Strobilomyces strobilaceus	3
	Tricholoma sejunctum	2
	Tricholoma squarrulosum	8
	Tulostoma brumale	1
Group D	*Amanita caesarea*	39
	Boletus torosus	8
	Clavariadelphus truncatus	2
	Cortinarius cedretorum	3
	Geastrum triplex	8

[a]Ing (1993).

storage purposes because they accelerate fruit body fermentation. Actions promoting good forest management and restricting grazing were also carried out.

A fundamental requirement is also to set a limit on the amount of mushrooms that may be picked by amateur and professional collectors. The rules also include that some species, such as *Boletus edulis* and related boletes, together with *Calocybe gambosa* and *Cantharellus cibarius* should not be collected if their caps are less than 2–3 cm in size. The picking of Caesar's mushroom (*Amanita caesarea*) in the early stages of development will also be forbidden in order predominantly to safeguard public health but also as a conservation measure.

Finally, a specific protection plan is proposed for *Pleurotus nebrodensis*, a rare Sicilian spring taxon with isolated distribution in the Madonie mountains of northern Sicily and the Etna mountain of eastern Sicily. The growing sites for *P. nebrodensis* are in the A zone (integral reserve) of the Madonie and Etna regional parks between 1400 and 2000 m, in meso-xerophilous pastures with *Cachrys ferulacea*. Since the seventeenth century, *P. nebrodensis* has been in great demand in the Madonie area because of its peculiar organoleptic characters. Because of the decrease in its natural fruiting, demand by consumers and restaurants exceeds the supply and results in high prices that vary from ITL 80 000 to 100 000 per kg. The increase in the number of collectors picking immature fruit bodies caused a conspicuous decline in fruiting and raised the risk of irreparable damage to the sustainability of this particular taxon. The protection plan for *P. nebrodensis* deals with legislative actions in order strictly to forbid all picking of fruit bodies throughout the Madonie area and to include it in the Italian Red Data list for fungi, as previously proposed by Venturella *et al.* (1997). In parallel to these safeguarding actions on this taxon, its *ex situ* cultivation was started (Venturella & Ferri, 1996; Venturella, 1999) to safeguard natural fruiting and as a sustainable alternative for the economy of the Madonie area.

Acknowledgements

The financial support of Ente Parco delle Madonie is gratefully acknowledged.

References

Ing, B. (1993). Towards a red list of endangered European macrofungi. In *Fungi of Europe: Investigation, Recording and Conservation* (ed. D. N. Pegler, L.

Boddy, B. Ing & P. M. Kirk), pp. 231–237. Royal Botanical Gardens: Kew, UK.

La Rocca, S. & Anastase, A. (1997*a*). Un interessante gasteromicete agaricoide in Sicilia: *Setchelliogaster tenuipes* (Setchell.) Pouzar. *Atti 4e Giornate Confederatio Europaea Micologiae Mediterraneae, Poggibonsi* (Italia), pp. 21–22.

La Rocca, S. & Anastase, A. (1997*b*). *Myriostoma coliforme* in Sicilia. *Micologia e Vegetazione Mediterranea* **12** (2), 114–120.

Venturella, G. (1992). Progetto per una banca dati sulla micoflora siciliana. *Quaderni di Botanica Ambientale e Applicata* **2** (1991), 107–110.

Venturella, G. (1995). *Tricholoma tridentinum* var. *cedretorum* Bon (*Tricholomataceae*), a misappreciated taxon from Sicily. *Documents Mycologiques* **98–100**, 465–467.

Venturella, G. (1999). The conservation of *Pleurotus nebrodensis* through its cultivation. In *Abstracts, XIII Congress of European Mycologists*, p. 131. Alcalà de Henares, Madrid, Spain, 21–25 September 1999.

Venturella, G. & Ferri, F. (1996). Preliminary results of *ex situ* cultivation tests on *Pleurotus nebrodensis*. *Quaderni di Botanica Ambientale e Applicata* **5** (1994), 61–65.

Venturella, G. & Mazzola, P. (1991). Present state of the mycological exploration in Sicily. 6th OPTIMA Meeting's Proceedings. *Botanika Chronika* **10**, 889–894, Delphi, 1989.

Venturella, G., Perini, C., Barluzzi, C., Pacioni, G., Bernicchia, A., Padovan, F., Quadraccia, L. & Onofri, S. (1997). Towards a Red Data List of fungi in Italy. *Bocconea* **5**, 867–872.

14
Fungal conservation in Ukraine

D. W. MINTER

Introduction

In countries that made up the former Soviet Union, however much one wants to concentrate on fungal conservation, general infrastructure problems demand attention. Until they have been addressed, little conservation of anything can take place. Ukraine is no exception. A number of agencies are committed to eliminating poverty and sustaining development in the transition countries in Central and Eastern Europe; the Department for International Development (DFID) is the British government department responsible for this. DFID has published a concise assessment of Ukraine's political, economic and environmental background which can be found in the UK Government's 1998 country strategy paper for Ukraine at the web site <http://www.dfid.gov.uk/public/what/pdf/ukraine_csp.pdf>, with the UK Government's environmental strategy for Ukraine being on <http://bc.kiev.ua/english/work/envstr.pdf>. The realities described in these two straight-talking documents make one admire all the more the dedication of those Ukrainian mycologists who have stayed in Ukraine and who somehow manage to continue their work in surroundings that are often difficult. Problems in Ukrainian science and education are a microcosm of those at national level. The issues that must be faced in solving those problems are extremely difficult, and many are contentious. It is inappropriate to discuss them here.

Ukraine does, however, have a national policy on nature conservation. The present work describes efforts by a team of Ukrainian and British scientists to provide infrastructure for fungal conservation within that national policy, through improved computing and informational resources. We have been able to organise a programme for the supply of second-hand computers. These are suitable for basic data entry of text information, but have severe limitations in more sophisticated use. Modern machines are needed for work with graphics, and for internet and modern e-mail connections. They need to be equipped with modems and CD-ROM drives, including at least some CD-ROM writers, and should be accompanied by some scanners and printers of a reasonable quality. Without at least some machines of that quality, fungal conservation in the former Soviet Union will always be behind the times. That sort of equipment, however, generally has to be purchased new. At present two major awards, from the UK Darwin Initiative (Biodiversity Information in the Former Soviet Union, 1999–2002), and the European Union organisation INTAS (Infrastructure Action for the Life Sciences, 2000–2002), and a smaller third award from the Royal Society of London (Electronic Distribution Maps of Ukrainian Fungi and Plants, 1998–2000) are helping to provide such equipment for critical institutions working with neglected groups of organisms, including fungi, in Ukraine and other newly independent states. The two major awards are also helping to pay for some e-mail connections and access to the Internet.

Mycologists in Ukraine

Ukraine has a great mycological tradition. The *Flora of Fungi of Ukraine*, the *Handbook of the Fungi of Ukraine* and various other major works document the mycological exploration of the country (Dudka & Burdiukova, 1996; Heluta, 1989; Kopachevskaya, 1986; Makarevich, Navrotskaya & Yudina, 1982; Merezhko, 1980; Merezhko & Smik, 1991; Morochkovskiĭ *et al.*, 1967, 1969, 1971; Oxner, 1937, 1956, 1968, 1993; Smik, 1980; Smitskaya, 1980, 1991; Smitskaya, Smik & Merezhko, 1986; Wasser, 1980, 1992; Zerova *et al.*, 1971; Zerova, Radzievskiĭ & Shevchenko, 1972; Zerova, Sosin & Rozhenko, 1979).

During the Soviet period, contact between Ukrainian scientists and their western counterparts was often difficult. There was a lack of money for travel and collaborative work, with the limited resources being allocated to senior scientists and people with good contacts in Moscow. Although the political barrier has gone, some of the other factors remain, so that a residual isolation still exists. It is timely, therefore, that the forthcoming 14th Congress of European Mycologists (14CEM) will be held in Crimea in 2003.

Computing resources

Conservation of biodiversity involves decisions and good decision-making requires accurate and readily available information. Nowadays, computers service this requirement. In the former Soviet Union computing resources available to people working with nature conservation vary enormously. The Zoology Institute of the Russian Academy of Sciences in St Petersburg has extensive computing skills and at least some modern computers. Some other institutions, for example biological research centres formerly with military significance, and some institutes and departments for experimental biology also have considerable facilities and skills. On the whole, though, the range of resources is highly skewed. Most biological institutions, universities, polytechnics and the like are very poorly equipped. The same is largely true for scientific societies, amateur societies (where they exist) and nongovernmental organisations, and even more so for schools, nature reserves, national parks, national nature parks, and private individuals with an interest in conservation. Some have no computers at all. This picture is general throughout the former Soviet Union. Ukraine is probably no better off than many other republics and, within Ukraine, mycology has often fared worse than other sciences.

Collecting and handling information about Ukrainian fungi

With a programme in place delivering computers to Ukraine for fungal conservation, the details of keyboarding, storing and using the desired information can be addressed. For fungi, biological records stem from four main sources: published records, unpublished lists, labels of reference collections, and field observations not backed by any specimens. To handle this information, three main databases are needed: a Nomenclatural and Taxonomic database, a Bibliographic database, and a database of the Biological Records themselves.

Problems with alphabets and instability of data

The first and most obvious problem when working in the former Soviet Union is one of alphabets. Scientific names, their authors, and places of publication of those names are essential components of any information system dealing with biological recording. These components, by and large, use the Latin alphabet. In the former Soviet Union, however, the Cyrillic alphabet is in general use for vernacular names of organisms, locations, names of collectors and identifiers, substrata, notes and many other aspects of biological recording. This means all keyboarding work needs ready and convenient access to both alphabets. Indeed, one quickly learns that the Cyrillic alphabet, like the Latin alphabet, exists in several forms: Russian, Ukrainian or Belarussian versions may be required at different times. Keyboarding is made much slower and more difficult by the constant need to switch between alphabets.

Anyone working with biological recording of neglected groups soon understands how labile information can be. The information system must accommodate, for example, the rapid changes in taxonomy which characterise contemporary mycology. This is done by keeping an unchanged copy of the original scientific name used as an identification of the organism in a separate field from current opinion of what that organism should be called. Biological databases that do not provide this basic service are of very doubtful value. Even so, handling data from the former Soviet Union can be a shock. Not only organism names but also, seemingly, every other item of data is prone to change. Within one human life span, for example, St Petersburg has been renamed to Petrograd, Leningrad and back again to St Petersburg. How many other place names in the former Soviet Union have also undergone name changes in the last ten years? New countries have appeared which did not previously exist. Regions and cities, towns and villages have changed their names. Vast numbers of 'Lenin Streets' have been swept away, together with 'Comsomol Avenues' and 'October Revolution Squares'.

The latter reminds us that the October Revolution itself occurred in November: an overdue calendar reform in the early 1920s changed the date, and thereby incidentally also changed the meaning of dates on every herbarium packet already in existence in every former Soviet Union botanical collection.

The inescapable conclusion is that, for proper data recording, particularly in the former Soviet Union, an intact copy of all original data should be stored in a suite of fields totally separate from those storing current opinion of what that information means. Filling in fields storing current information is, of course, a different job from data gathering. It is a large overlay, ideally carried out by trained and experienced editors. There is also, however, a further complication. If the only objective for using the resulting information is domestic, records with the organism name in the Latin alphabet, and all other information in the Cyrillic alphabet may be adequate. But if the longer-term objective is international use of the data, then a second copy of all of the information, all in the Latin alphabet, is essential. This means that, for all biological records collected in the former Soviet Union, the database system must have, at least for the critical data elements, three copies of all information: first, the original data; second, a modern opinion of what that information means, stored in the language of the owner country; third, an English language translation of that modern opinion. In all the collaborative work done by my colleagues in Ukraine, this daunting challenge has been faced.

Sources of information and information storage

At an early stage in collaboration, in 1994, a decision was made to use a Nomenclatural & Taxonomic database already in existence at the former International Mycological Institute (now CABI Bioscience). In this way, a large amount of duplication of effort was avoided. The Nomenclatural & Taxonomic database then stored about 30 000 different scientific names of organisms. The number has subsequently grown to a current figure of over 450 000 names including not only fungi, but also plants, bacteria, protozoans and animals (mainly insects). The system is designed to cope easily with all taxonomic ranks, and has a built-in warning system to alert the user to ambiguities arising from homonyms and orthographic errors. In most cases, the information needed to connect to the appropriate record in this database is simply the Latin name of the organism, stripped of any accents or hyphens.

The Nomenclature & Taxonomic database is itself linked to a Bibliographic database then containing about 3000 (now more than 40 000) records of different published works on mycology, and it was therefore natural to decide to use that database too. There were, however, problems. That Bibliographic database, having been developed in western Europe, was (and still is) rather weak in its coverage of Cyrillic alphabet publications. Furthermore, those Cyrillic language records that it holds generally only contain either an English translation or a Latin alphabet transliteration of the original data. A small project was therefore undertaken from 1995 to 1996, generously funded by the British Council in Kiev, to establish a new Bibliographic database in Ukraine specifically for mycological publications in the Cyrillic alphabet. That database is structurally identical to the main Bibliographic database at CABI Bioscience, and in due course a copy of the Cyrillic records will be shared by both systems. As a first stage, all the bibliographic information about mycological publications in *Mikologiya i Fitopatologiya* and *Ukrainian Botanical Journal* is being input.

A fully structured database for Biological Records is very complex, and involves many different tables (Minter, 1996). Such a database is suitable for long-term storage of data, but is rarely appropriate for the data gathering stage. For that work, far simpler customised data-entry schemes are needed. For collaboration in Ukraine, a feeder database intermediate in complexity and specially designed as a general data-entry system was adopted. Such a database permits rapid keyboarding, while ensuring that the critical data elements are separated, thus minimising the later editorial overhead (a range of computer programs exists at CABI Bioscience to carry out mechanically all the commoner editorial jobs). Copies of that feeder database were first installed in Kiev in February 1994, with the first donated computers to reach mycologists in Ukraine. Installation of that system marked the beginning of an intensive phase of computerising records of fungi in Ukraine, which continues at the time of writing.

A computerised checklist of the fungi of Ukraine

The priority of the first collaborative work, a UK Darwin Initiative project (*Fungi of Ukraine*, 1993–1996), was to computerise the information on herbarium labels in fungal collections of the *M. G. Kholodny Institute of Botany* in Kiev. The initial target was to digitise 24 000 records in 27 months. In the event, over 48 000 records of fungi were keyboarded, an impressive total, given the difficulties of setting up the system, and of abstracting and keyboarding often almost illegible data hand-written in Cyrillic on frequently faded packets. Many of these 48 000 records were of fungi occurring on a host plant, or some other associated organism. In the fully structured database for Biological Records, these observations of associated organisms are treated as separate linked records valid in their own right. The result was that, after editing, another approximately 32 000 records of associated organisms were added to the original 48 000 fungal records, making a total of around 80 000 biological records of all groups of organisms from Ukraine generated by that project. These were used to produce Ukraine's first computerised checklist of its fungi (Minter & Dudka, 1996).

Generation of new data

It is always important to maintain a continued flow of new observations in addition to computerising existing records. The UK Darwin Initiative project, the Royal Geographical Society's first Ralph Brown Prize (for an expedition to the Pripyat Marshes of Ukraine) and awards from the British Council and the UK Know How Fund have helped to keep fieldwork in Ukraine alive. British Council funding permitted British and Ukrainian scientists to visit Crimea in 1993. Through the Darwin Initiative project, various collecting trips were organised for teams of Ukrainian scientists, including the Carpathian Mountains (1994, 1995), L'viv Oblast (1995) and Crimea (1996). The British Mycological Society paid for one Ukrainian student to attend their Polish Foray in 1995 (incidentally providing an opportunity to transport back-numbers of *Mycological Research* and *The Mycologist* donated by the Society to Ukraine at the same time). Know How Fund support enabled British and Ukrainian scientists to visit the Ukrainian Steppe Reserve in Donetsk Oblast in 1997. The Royal Geographical Society Ralph Brown Expedition put up to fifty scientists and students of a wide range of disciplines into the field for periods of up to two months during summer 1998. One result was over one hundred species of fungi new for Ukraine. Realities of life in Ukraine mean that much fieldwork has to be done under canvas. These awards, and particularly the Ralph Brown Prize, provided much important field equipment to Ukrainian scientists – cagoules, boots, overtrousers, tents, sleeping bags, airbeds, cooking equipment, canoes and many other items – without which future fieldwork would be greatly hampered.

Electronic distribution maps of Ukrainian fungi

After the 1994–1996 UK Darwin Initiative project ended, development of the computerised databases continued. The Royal Society project (Electronic Distribution Maps of Ukrainian Fungi and Plants, 1998–2000) enabled mycologists in Ukraine to add latitude and longitude co-ordinates to all records for which locality information was sufficiently precise (over 35 000). For place names not included in the various available gazetteers this was no small task. The best maps readily available in Ukraine do not show latitude and longitude, and have place names in Russian rather than Ukrainian. It was frequently necessary to locate a place on these maps, and then calculate the location by reference to another more general map that did show latitude and longitude. Under these difficult conditions more than normal care was needed to ensure accuracy of data. In collaboration with Dr A. H. Thomas at the NERC Institute of Terrestrial Ecology, Bangor, simple electronic distribution maps were then produced of a sample 75 species, and these have been made available on the Internet (<http://www.biodiversity.ac.psiweb.com/>).

Greatly extended computerisation of fungal records

With the new UK Darwin Initiative award (Biodiversity Information in the Former Soviet Union, 1999–2002), it has been possible to extend the work computerising the fungal collections in the herbarium of the M. G. Kholodny Institute of Botany. Attention is now being directed to the larger basidiomycetes which, for various operational reasons, could not be prioritised during the first project. Work is also beginning to computerise fungal records in the reference collection of the Nikita Botanic Garden (Yalta), using this opportunity to issue proper accessions numbers for the first time. Now that their bibliographic information has been computerised, work is also under way extracting records of fungi from *Mikologiya i Fitopatologiya* and *Ukrainian Botanical Journal*. For this work, a policy decision was made to extract all records, and not just those from Ukraine. To date, about 7000 fungal records have been computerised from these sources, but the final total is likely to run to several tens of thousands. When this work is complete there will be an enormous resource of information about fungi from all parts of the former Soviet Union.

The new UK Darwin Initiative project is also, incidentally, permitting our team to extend computerised biodiversity coverage in other ways, some reaching beyond Ukraine. The already published checklist of the fungi of Georgia (Nakhutsrishvili, 1986) is being computerised, with nearly 17 000 records already keyboarded, and collaborative work has begun computerising mycological herbaria in Russia and Kazakhstan. During 2000, Ukrainian mycologists expect to help Georgian and Armenian mycologists to begin computerising fungal collections in the Botanical Institutes of Tblisi and Yerevan. Finally, and important for Ukrainian fungal conservation, this project is helping with the publication of a new edition of the computerised checklist of Ukrainian flowering plants.

A new directory of scientists

Anyone wishing to improve infrastructure in the former Soviet Union is spoilt for choice. The trick is to select a problem that can be resolved within a practical time frame. At present there is no up-to-date directory listing names and addresses of people and institutions working on biodiversity and nature conservation in the former Soviet Union. Such a directory is particularly necessary because the ten years since the fall of the Soviet Union have been chaotic: new countries have come into existence, and with them new governments, new ministries, new academies and new administrative systems; many cities, towns, villages and streets have changed their names; people have emigrated or moved in response to economic and social pressures, and sometimes even to war; new technologies, such as fax, e-mail and the Internet have become widespread. The old Soviet directories are no longer usable. This is one of the biggest infrastructure problems for this area of science, but it is also a problem that can be resolved. Ukrainian mycologists at the M. G. Kholodny Institute of Botany, in collaboration with biologists at the N. K. Koltsov Institute of Developmental Biology in Moscow, are playing a major role in the development of a new computerised directory.

This work is funded by awards from the UK Darwin Initiative (Biodiversity Information in the Former Soviet Union, 1999–2002), and the European Union organisation INTAS (Infrastructure Action for the Life Sciences, 2000–2002). The first version of the new directory will use the Latin alphabet (English language), with the aim of making it easier for outsiders interested in biodiversity and nature conservation to establish contact with key people in former Soviet Union countries. Subsequently one or more versions of the directory using the Cyrillic alphabet will be distributed within the former Soviet Union, to enhance internal contacts. A meeting will be held in 2002, probably in Ukraine, under the auspices of BioNET INTERNATIONAL. That meeting will draw together key people, particularly those working on neglected groups such as invertebrates, fungi and micro-organisms, with the aim of establishing a new general infrastructure for biodiversity and nature conservation work within the former Soviet Union.

14th Congress of European Mycologists

The acceptance of Ukraine's bid to host 14CEM in Yalta (Crimea) in 2003 is a major boost to the country's mycologists. It will help them enormously in developing infrastructure for mycology in Ukraine and, with the traditional emphasis of the Congress on conservation, they will be stimulated to formulate a new national strategy for fungal conservation. The congress will put Ukraine at the centre of European mycology for three years, at a time when fungal mapping and fungal conservation at a European level are getting off the ground, and when a team appointed by the 13th Congress is looking at the possibility of a European Mycological Association. The influx of substantial numbers of western European mycologists to the congress will permit mycologists from Ukraine (and other former Soviet Union countries) to establish new collaborative networks, while the organisation of the congress itself will require development of new skills with e-mail and the Internet, and a much greater facility with western European languages. Furthermore, the western mycologists visiting Crimea will gain new insights into the mycological treasures so worth conserving in Ukraine. The biggest challenge will be to find the funding necessary to bring in the new computers, printers, photocopiers, modems and other Internet paraphernalia which the organisation of this congress will require.

Plans for the future

As Ukrainian mycologists adapt and survive in the new post-Soviet environment, a trend can be detected for them to become more adventurous. The model of Spain, an isolated country on the death of Franco but now fully integrated into Europe, and the impressive achievements of Spanish mycologists provide them with hope that Ukraine's larger infrastructure problems will be resolved, and with inspiration that at the same time they can grow. They realise that broader issues can now be identified and tackled, because of the solid informational base which now exists, and because they have considerably more experience of collaborating with westerners and approaching donor organisations. This means the range of areas in which they are prepared to work to conserve fungi is increasing, and there is lively debate about which problems are now the highest priority: their ideas are taking on the character of a manifesto.

As everywhere in the world, the loss of habitat is probably the most significant threat to fungi in Ukraine, and again that means broader infrastructure issues. At the time of writing, the team is involved with various proposals, including: emergency work to reduce catastrophic flooding in Volyn Oblast (based on experience and information gained during the Ralph Brown expedition); introduction of controlled grazing to prevent reversion of steppe to scrub (steppe is one of the most endangered habitats in Ukraine); a publicity campaign to discourage illegal mass gathering and selling of wild flowers. It seems likely that there will also be future interest in at least some of the following measures: developing a specialist industry in sustainable tourism (an award from the British Council is currently helping part of the team to look at possibilities in this area); protecting Crimea's beautiful steppe meadows, monitoring the effects of new housing schemes; the damage done to fungi by pollution; the role of fungi in pollution monitoring, and bioremediation of pollution (particularly relevant with regard to radioactive pollution after Chernobyl).

As more computers reach different biosphere reserves, nature reserves and national nature parks, there will be also an increased emphasis on helping those organisations to develop management plans that take into account conservation of the fungi. In this work, Ukrainian mycologists will be looking to their western colleagues from whom they will need considerable support. They will also try to use their computerised data and other information to identify sites of special scientific interest for mycology, and will seek with other biologists to encourage the Ukrainian military to recognise the conservation value of some of their training grounds. Funds will be sought to help the country's two great fungal culture collections (in the M. G. Kholodny Institute of Botany, and the D. K. Zabolotny Institute of Microbiology and Virology) to survive and grow. Every effort will be made to ensure that conservation nongovernmental organisations, like the National Ecocentre of Ukraine, are supported and in turn support and publicise fungal conservation. In the midst of these simultaneously daunting and exciting opportunities, two particular courses of action are receiving detailed attention.

Red lists, grey lists, green lists

As information becomes available in computerised format, many exciting prospects are opening up for fungal conservation in Ukraine. It is already possible to assess Ukraine's known fungi by taxonomy, to identify which groups are under-recorded. It is possible to pick out all taxa with fewer than a specified number of records, or taxa with a most recent date of observation earlier than, say, forty years ago. Such mechanical analyses are essential in preparing new red data lists according to IUCN criteria, but they can also be used as a tool to show the inadequacies of those criteria, developed for the charismatic megafauna, when applied to fungi. There is currently no red data book in Ukraine for fungi, but a few fungi are present in the current edition of the red data book of Ukraine for plants (Shelyag-Sosonko, 1996), almost all larger basidiomycetes. Now, however, mycologists in Ukraine have a real chance to produce a meaningful red data list of their fungi for the first time.

That list is unlikely to be a book, and probably won't be a single list either. It is more likely to take the form of a CD-ROM and perhaps a website, with several different lists. These will include the conventional red list of fungi known to be endangered, but if the IUCN criteria are used, that list may contain so many presumed extinctions – species not observed for more than 40 years – that some additional notation would be necessary. There will also be a list of fungi recorded in Ukraine fewer than, say, 20 times, and lists of species known to be limited in distribution. The existence of fully computerised checklists for Ukraine of both fungi and flowering plants, together with computerised information about fungal plant associations and substrata, however, also permits the production of lists of Ukrainian plants on which no fungi have been observed, and lists of Ukrainian plants for which, for example, fungi have only been observed on leaves, but not on twigs, branches, trunks, wood, roots, flowers etc. These lists – *grey* lists of the unknown rather than *red* lists of the known – will for the first time allow reasonably precise indications about what remains to be explored. Ukraine is unusual in that it already has lists (known within Ukraine as 'Green Lists') of endangered plant communities. Since fungal conservation is so tied to habitat conservation, these plant community lists will be another important factor in future in assessing the status of the country's fungi.

A proposal is now in preparation to produce that CD-ROM in time for 14CEM. In addition to the various lists, the disk would also contain over 5000 distribution maps, one for every fungal species known from Ukraine, building on the earlier Royal Society project. Although in most cases the amount of information is insufficient for a real concept of the species' geographical distribution, this product will provide a base line for future recording work. Work is also beginning to develop one or more associated websites, to raise the profile of Ukrainian fungal conservation internationally. Ukrainian mycologists are aware that 'Red Lists', of whatever colour, are political documents, and care will be taken to ensure that any resulting CD-ROMs will produce the best possible political outcomes. One of the targets may be a revised Red Data Law for Ukraine taking more account of the need to conserve fungi.

A mycological society for Ukraine?

There are various scientific societies in Ukraine dealing with different groups of organisms. Comparison of these societies with analogous bodies in, for example, the European Union is instructive. Scientists from outside Ukraine might find the practice in certain societies of having the same person president for many years with no bar on re-election a little unusual. In the west a regular change at the top is considered desirable to prevent stagnation and the development of undesirable patronage. There is no scientific society in Ukraine specifically for the study of fungi. Mycologists (including lichenologists) tend to be members of the Ukrainian Botanical

Society, which has a section devoted to Plant Pathology. With the fungi now universally accepted as having at least one kingdom of their own, perhaps the time is coming to establish a new and separate mycological society in Ukraine, run on modern lines, and free from the historical constraints which continue to affect other already existing scientific societies in Ukraine.

Such a society could play an important role in fungal conservation, even with limited funds. This could be realised through the development of field meetings, workshops, lectures, displays, fostering of local groups, recording and mapping schemes, an educational programme (including pressure to raise the profile of mycology in schools and tertiary education), posters, publicity, modernised publications, organised fund-raising (perhaps through an encouraged amateur membership), websites and all the other tools so successfully used by other learned societies outside Ukraine. A new mycological society for Ukraine could also act as a unified voice for Ukrainian mycology within the European and international scientific communities, and would probably be able to attract significant grants at establishment stage from European Union bodies fostering the development of democratic institutions within former Soviet Union countries.

Acknowledgements

None of the work for fungal conservation in Ukraine described in this chapter would have been possible without generous support from a range of people and organisations. The National Academy of Sciences of Ukraine has always supported our work directly or indirectly, particularly through the M. G. Kholodny Institute of Botany, the country's principal institution for work on fungal biodiversity and conservation and its Director, Academician K. M. Sytnik. Awards from the UK Darwin Initiative, the European Union organisation INTAS, the Royal Geographical Society's Ralph Brown Award, the British Council in Kiev, the Royal Society of London, the UK Know How Fund and the British Mycological Society have provided the financial foundations for the work. Donations of computing and other equipment, particularly from CAB International, the NERC Institute of Terrestrial Ecology Bangor, Richard Ellis & Co. and the National Trust, and assistance in transporting that equipment from SAS, British Airways, Ukrainian International Airlines and Zoloti Kolesa have made the seemingly impossible seem easy. The staff of the Embassy of Ukraine in London is thanked for their kind co-operation. The enthusiasm of all members of the 'Darwin Team' in Kiev, and their families, is gratefully acknowledged. Lastly the author expresses his gratitude to his employers for their support, and to his family for their toleration of the piles of second-hand computers which have periodically invaded the house.

References

Dudka, I. A. & Burdiukova, L. I. (1996). Oomitsety: Fitoftorovye i albugovye griby [Oomycetes: Phytophthoraceae Pethybr. and Albuginaceae Schröt.]. *Flora Gribov Ukrainy* [*Flora of the Fungi of Ukraine*], 208 pp. Naukova Dumka: Kiev.

Heluta, V. P. (1989). Muchnistorotsyanye griby [Powdery mildew fungi]. *Flora Gribov Ukrainy* [*Flora of the Fungi of Ukraine*], 256 pp. Naukova Dumka: Kiev.

Kopachevskaya, E. H. (1986). *Likhenoflora Kryma i yeye Analiz* [*The Lichen Flora of Crimea and its Evaluation*], 296 pp. Naukova Dumka: Kiev.

Makarevich, M. F., Navrotskaya, I. L. & Yudina, I. V. (1982). *Atlas Geograficheskogo Rasprostraneniya Lishainikov v Ukrainskikh Karpatakh* [*Geographical Atlas of the Distribution of Lichens of the Ukrainian Carpathians*], 403 pp. Naukova Dumka: Kiev.

Merezhko, T. A. (1980). Sferopsidal'ny e griby [Sphaeropsidalean fungi]. *Flora Gribov Ukrainy* [*Flora of the Fungi of Ukraine*], 208 pp. Naukova Dumka: Kiev.

Merezhko, T. A. & Smik, L. V. (1991). Diaportal'nye griby [Diaporthalean fungi]. *Flora Gribov Ukrainy* [*Flora of the Fungi of Ukraine*], 216 pp. Naukova Dumka: Kiev.

Minter, D. W. (1996). Recording and mapping fungi. In *A Century of Mycology* (ed. B. C. Sutton), pp. 321–382. Cambridge University Press: Cambridge.

Minter, D. W. & Dudka, I. O. (1996). *Fungi of Ukraine, a Preliminary Checklist*, 361 pp. International Mycological Institute: Egham, & M. G. Kholodny Institute of Botany: Kiev.

Morochkovskiĭ, S. F., Radzievskiĭ, G. G., Zerova, M. Ya., Dudka, I. O., Smits'ka, M. F. & Rozhenko, G. L. (1971). Nezaversheny gribi [Imperfect fungi]. *Visnachnik Hribyv Ukraini* [*Handbook of the Fungi of Ukraine*] **3**, 1–696. Naukova Dumka: Kiev.

Morochkovskiĭ, S. F., Zerova, M. Ya., Dudka, I. O., Radzievskiĭ, G. G. & Smits'ka, M. F. (1967). Slizoviki (Myxophyta), Hribi (Mycophita): Arkhymytseti, Fykomytseti [Slime moulds (Myxophyta); Fungi (Mycophyta): Archymycetes, Phycomycetes]. *Visnachnik Hribyv Ukraini* [*Handbook of the Fungi of Ukraine*] **1**, 1–255. Naukova Dumka: Kiev.

Morochkovskiĭ, S. F.; Zerova, M. Ya.; Lavitskaya, Z. G.; Smitskaya, M. F. (1969). Askomytseti [Ascomycetes]. *Visnachnik Hribyv Ukraini* [*Handbook of the Fungi of Ukraine*] **2**, 1–518. Naukova Dumka: Kiev.

Nakhutsrishvili, I. G. (1986). *Flora Sporovykh Rasteniy Gruzii* [*Flora of Sporulating Plants of Georgia*], 888 pp. Metsniereba: Tblisi.

Oxner, A. M. (1937). *Viznachnik Lishainikyv URSR* [*Handbook of Lichens of the Ukrainian SSR*], 341 pp. Institute of Botany, Academy of Sciences of the Ukrainian SSR: Kiev.

Oxner, A. M. (1956). *Flora Lishainikyv Ukraini* [*Flora of the Lichens of Ukraine*] **1**, 1–495. Naukova Dumka: Kiev.

Oxner, A. M. (1968). *Flora Lishainikyv Ukraini* [*Flora of the Lichens of Ukraine*] **2** (1), 1–500. Naukova Dumka: Kiev.

Oxner, A. M. (1993). *Flora Lishainikyv Ukraini* [*Flora of the Lichens of Ukraine*] **2** (2), 1–544. Naukova Dumka: Kiev.

Shelyag-Sosonko, Y. R. (1996). *Chervona Kniga Ukraini: Roslinniy Svit* [*Red Book of Ukraine: Plant Kingdom*]. Edition 2, 668 pp. Vidavnitvo 'Ukrains'ka Entsiklopediya' im. M.P. Bazhana: Kiev.

Smik, L. V. (1980). Sferial'nye griby [Sphaerialean fungi]. *Flora Gribov Ukrainy* [*Flora of the Fungi of Ukraine*], 184 pp. Naukova Dumka: Kiev.

Smitskaya, M. F. (1980). Operkulyatnye diskomitsety [Operculate discomycetes]. *Flora Gribov Ukrainy* [*Flora of the Fungi of Ukraine*], 224 pp. Naukova Dumka: Kiev.

Smitskaya, M. F. (1991). Gipokreal'nye griby [Hypocrealean fungi]. *Flora Gribov Ukrainy* [*Flora of the Fungi of Ukraine*], 89 pp. Naukova Dumka: Kiev.

Smitskaya, M. F., Smik, L. V. & Merezhko, T. A. (1986). Opredelitel' pirenomitsetov USSR [*Handbook of the Pyrenomycetes of the Ukrainian SSR*], 364 pp. Naukova Dumka: Kiev.

Wasser, S. P. (1980). Basidiomitsety. Agarikovye griby [Basidiomycetes: Agaricalean fungi]. *Flora Gribov Ukrainy* [*Flora of the Fungi of Ukraine*], 326 pp. Naukova Dumka: Kiev.

Wasser, S. P. (1992). Basidiomitsety. Amanital'nye gryby [Basidiomycetes: Amanitalean fungi]. *Flora Gribov Ukrainy* [*Flora of the Fungi of Ukraine*], 186 pp. Naukova Dumka: Kiev.

Zerova, M. Y., Morochkovskiĭ, S. F., Radzievskiĭ, G. G. & Smits'ka, M. F. (1971). Basidyomytseti: Dakrymitsetal'ni, Tremelal'ni, Aurykularyal'ni, Sazhkovn'dni, Irzhasti [Basidiomycetes: Dacrymycetales, Tremelales, Auriculariales, Ustilaginales, Uredinales]. *Visnachnik Hribyv Ukraini* [*Handbook of the Fungi of Ukraine*] **4**, 1–316. Naukova Dumka: Kiev.

Zerova, M. Y., Radzievskiĭ, G. G. & Shevchenko, S. V. (1972). Basidyomytseti: Eksobasydyal'ni, Afyloforal'ni, Kanteral'ni [Basidiomycetes: Exobasidiales, Aphyllophorales, Cantharellales]. *Visnachnik Hribyv Ukraini* [*Handbook of the Fungi of Ukraine*] **5** (1), 1–240. Naukova Dumka: Kiev.

Zerova, M. Y., Sosin, P. E. & Rozhenko, G. L. (1979). Basidyomytseti: Boletal'ni, Strobilomitsetal'ni, Trikholomatal'ni, Entolomatal'ni, Russulal'ni, Agarykal'ni, Gasteromitseti [Basidiomycetes: Boletales, Strobilomycetales, Tricholomatales, Entolomatales, Russulales, Agaricales, Gasteromycetes]. *Visnachnik Hribyv Ukraini* [*Handbook of the Fungi of Ukraine*] **5** (2), 1–566. Naukova Dumka: Kiev.

15

The threatened and near-threatened Aphyllophorales of Finland

H. KOTIRANTA

Introduction

The total area of Finland is 337 000 km^2. The land area is 305 000 km^2 and 65.8% (about 200 000 km^2) of this is wooded. Most of the aphyllophoroid fungi are wood-decayers, saprotrophs or parasites, and only a handful can form mycorrhizas.

Altogether 848 aphyllophoroid taxa were evaluated according to the 'IUCN Red List Categories' (Anon., 1995), with a national application. It means that the numbers used in different criteria (see below) on the national level are smaller than in the original IUCN criteria.

Categories

The following categories were used: EX (Extinct), CR (Critically Endangered), EN (Endangered), VU (Vulnerable), NT (Near Threatened), DD (Data Deficient), LC (Least Concern) and NE (Not Evaluated). In the tables that follow, the species which are near threatened (NT) are considered also, even though they are not threatened (yet?) in the strict sense of the IUCN criteria.

Criteria

The criteria (which are fully described in Anon., 1995) that have been used in different threat classes are briefly as follows; they have been applied as recommended by the 40th meeting of the IUCN Council, except in class NT, which is a IUCN category for a population that has diminished by more than 10% during ten years or three generations:

Table 15.1. *Aphyllophoroid taxa in different threat categories*

	Categories							
	EX	CR	EN	VU	NT	DD	LC	NE
Aphyllophoroid taxa	5	22	31	62	75	171	478	4
Percentage value	0.6	2.6	3.7	7.3	8.8	20.1	56.4	0.5

A, reduction of population (observed, estimated, inferred or suspected);
B, distribution area (small, strongly fragmented, etc.);
C, small and continuously declining population;
D, very small and restricted population, which is then subdivided into subcategories on the basis of population size or distribution area thus: D1 comprises CR, less than 50 individuals; EN, less than 250 individuals; VU, less than 1000 individuals; and NT, less than 5000 individuals; whilst in D2 the distribution area of the population is not more than CR, $<5\,km^2$ or one area of occupancy only; EN, $<50\,km^2$ or two areas of occupancy only, VU, $<100\,km^2$ or 3–5 areas of occupancy only; and NT, $<500\,km^2$ or 25 areas of occupancy only.

Results

Table 15.1 shows that the threatened categories, EX, CR, EN and VU, together comprised 120 species, which represented 14% of the total taxa. The few species that were 'Not Evaluated' (category NE) all grow inside some sort of building, such as *Serpula lacrymans* (in dwellings) or *Gloeophyllum trabeum* (found in saunas), and are not found in natural habitats in Finland. The group described as Data Deficient (DD) is relatively large. There are tens of species which have been found once or only a few times. Their right place, at least for the moment, is in class DD rather than in one of the threat classes, such as CR. Some newly described species are classified in DD also. In the 'Near Threatened' class (NT) are species which do not fulfil the usual criteria of vulnerable species. Many of the taxa in class NT are old-growth forest species.

Of the criteria used in defining the different classes (Table 15.2), criterion A (reduction of population) is often used in the less-threatened classes (VU, NT). These species have their optimum in virgin forests. The reduc-

Table 15.2. *The criteria used in different classes (note: all the criteria are tested, and several criteria per species are used if they fulfil them)*

IUCN criterion	% in each threat category			
	CR	EN	VU	NT
A (reduction of population)	22.7	38.7	45.2	53.3
B (distribution area)	40.9	25.8	46.8	5.3
C (small and continuously declining population)	13.6	35.5	12.9	4.0
D (very small and restricted population)	100	87.0	74.2	72.0

Table 15.3. *Frequency of instances in which only one criterion is fulfilled in different classes*

IUCN criterion	Proportion in each threat category			
	CR	EN	VU	NT
A (reduction of population)	0/22	0/31	11/62	19/75
B (distribution area)	0/22	0/31	1/62	0/75
C (small and continuously declining population)	0/22	0/31	0/62	0/75
D (very small and restricted population)	11/22	11/31	10/62	29/75

tion of the old-growth forests over three generations has greatly increased. In Finland, the age of a generation of most wood-rotting fungi that infest large trunks only is concluded to be 25 years, even longer in many cases.

Criterion D (very small and restricted population) is used in all of the Critically Endangered species (Table 15.2). The high percentage even in class NT shows that the listed species have small populations and the number of individuals is not very high. In this study one tree trunk which is occupied by a mycelium and bears fruit bodies of one species is considered to be one individual, regardless of how many fruit bodies there are on the trunk. This is, of course, not the true situation for every species (e.g. *Trametes* spp.), but it is the only practical way to solve the problem of defining an individual at the moment without extensive *in vitro* studies. The frequency with which only one criterion is fulfilled in the different classes is shown in Table 15.3.

Table 15.4 shows the primary threats in different classes. The greatest threats for the Finnish Aphyllophorales are the changes in age structure of the forests (reduction of virgin forests) and the reduction of decaying wood. It is noteworthy that none of the species is threatened because of

Table 15.4. *The primary threats in different classes*

Primary threat	% in each threat category				
	EX	CR	EN	VU	NT
Changes in age structure of forests/forest stands (reduction of primeval forests, giant trees and burnt areas)	0	41.0	48.4	46.8	52.0
Reduction of decaying wood (e.g. hollow or decayed trees)	60	27.3	29.0	24.2	26.6
Plantation and forest management activities	0	13.6	6.5	11.3	8.0
Changes in tree species composition (domination by spruce and reduction in deciduous tree stand)	40	13.6	0	8.1	8.0
Construction: towns, rural areas, shores, road building etc.	0	4.5	12.9	6.4	2.7
Construction of waterways	0	0	3.2	0	0
Peatland drainage for forestry and peat harvesting	0	0	0	1.6	0
Reason/cause unknown	0	0	0	1.6	2.7

Note: In very small populations the small size is already a serious threat, but it is used only if no other threats are known.

Table 15.5. *The primary habitats in different classes*

Primary habitat	% in each threat category				
	EX	CR	EN	VU	NT
Old heath forests	0	40.9	45.2	46.8	48.0
Heath forests	100	13.6	6.5	8.1	29.3
Old herb-rich forests	0	4.5	3.2	11.3	5.3
Herb-rich forests	0	36.4	32.2	29.0	13.3
Lake shores and river banks	0	0	9.7	1.7	0
Parks, yards and gardens	0	4.5	0	1.7	1.3

Note: Only the most common habitats are mentioned, habitats of one species only are excluded.

picking or collecting or by chemical disturbance like atmospheric pollution.

The overwhelming majority of threatened species (also those categorised as near threatened) are forest species and the most common habitat is old-growth heath forest (Table 15.5) with Norwegian spruce (*Picea abies*), pine (*Pinus sylvestris*), birches (*Betula pendula*, *B. pubescens*) and aspen trees (*Populus tremula*). The most important host is spruce, followed by pine and aspen.

Future

During the last ten years an intensive survey of old-growth forests took place in Finland. In the survey the most valuable areas were investigated. To evaluate the conservation value of forests several criteria were used, including the amount, quality and continuity of dead wood, and some polypores were used as indicators of primeval forests.

The survey led to protection of large areas of old-growth forests. Moreover, the National Parks, Strict Nature Reserves and areas otherwise protected by law nowadays form a good network of old-growth forests. The already existing protected areas and those that are in different nature conservation programmes (e.g. Natura 2000) now cover 714 300 ha. If less productive forests are also included (mires with pines etc.) the protected area is over 1.5 million ha. However, most of the protected areas are situated in the northern part of the country, and the area protected in southern Finland is still too small. The situation of the most threatened Aphyllophorales is now good in northern Finland, but needs further action in southern Finland.

Some of the species (mostly in the threat classes CR and EN) are classed as Specially Protected. If some of these species are found outside a protected area the management (forest cutting, road building etc.) of the area is interrupted until estimations of the management needed to protect the species is established.

The new forest cutting recommendations, that some amount of wood (often old, large or already dead trees) is always left in the managed area, may help some species to survive even in commercially managed forests.

Reference

Anon. (1995). *IUCN Red List Categories*. IUCN: Gland, Switzerland.

16

Fungal conservation in Cuba

D. W. MINTER

Introduction

Cuba, the largest island in the Caribbean, is in many ways a remarkable place. Its location gives it this special character. The country lies just inside the tropics, in the Greater Antilles archipelago, between the Dominican Republic and the Yucatan Peninsula of Mexico, with Florida to the north. The island's main crops are sugar cane and tobacco, and much of the land is devoted to agriculture. There are also, however, many areas of unspoilt and outstanding natural beauty. These include the mogotes (limestone hills) in the west, the huge expanses of the Ciénaga de Zapata mangrove swamps in the south, the Sierra Maestra range in the east, and numerous coral reefs and small islands around the main island. At its western end, the island's flora and fauna contain elements from North America, for example native oaks. Further east the mix of organisms gradually becomes truly Antillean. Cuba is, furthermore, sufficiently isolated from other landmasses to ensure that about 50% of its approximately 6500 native plant species are endemics, making it the region's principal centre of speciation.

The human population of about 11.2 million is of mixed racial origin, Spanish speaking and partly urbanised. The country's natural trading partner is the USA and, before 1959, Cuba was for many years virtually a colony of the US. Since the revolution, however, Fidel Castro's government has maintained a policy of genuine independence from the USA, with strong backing from the Soviet Union: his famous foreign minister, Che Guevara, organised the equally famous barter of Cuban sugar for Soviet oil. The USA responded with an economic blockade, which has lasted to today, and is still being tightened. The blockade is unilateral, and regularly opposed by most members of the United Nations, including the UK and other countries of the European Union. The revolution, while criticised in some quarters, resulted in phenomenal improvements to education and public health. Cuba now has a highly educated workforce, with low levels of infant mortality, excellent life expectancy statistics, and an infrastructure unparalleled in any other country with a similar level of GNP.

The collapse of the Soviet Union, however, removed an important economic prop and, combined with the blockade, resulted in significant hardship for the civilian population over the last decade. In response to that hardship, which was known in Cuba as 'the special period', various initiatives were launched, particularly in biotechnology and tourism. Certain economic adjustments were also introduced which removed some restrictions on the handling of foreign currency. These have resulted in considerable trade growth in recent years with Canada, Mexico and the European Union, and some easing of the economic hardship. They have also made some international scientific collaboration easier.

In terms of mycology Cuba is no less remarkable. The earliest fungal explorations were made in the mid nineteenth century by the British and French, followed by considerable work at the end of the nineteenth century and the beginning of the twentieth century by mycologists based in the USA. Large numbers of early and significant Cuban fungal specimens therefore reside in reference collections outside the country, particularly in North America. Over the last 40 years, the island has produced a string of world-class scientists, trained largely in the former East Germany and former Czechoslovakia but also in Canada and Russia. Their publications, and those of their mentors, particularly Kreisel (1971*a*,*b*) and Arnold (1986) have contributed greatly to our knowledge of Caribbean fungi. These mycologists have also organised impressive, large and fully functional herbaria and culture collections within Cuba, and it can be reasonably argued that no other country in Latin America is better equipped to carry out serious work on fungal biodiversity.

Under the very difficult conditions of the US blockade, the same scientists were instrumental in organising the first two Latin-American Congresses of Mycology, thereby effectively starting the Asociación Latinoamericana de Micología. My participation in the second of those congresses, in Havana in 1996, gave me the stimulus to propose a project through the UK Darwin Initiative with the objectives of preparing a computerised checklist of the fungi of the Caribbean, putting into place the infrastructure for a regional fungal identification service, and drawing up proposals for national fungal conservation strategies for two Caribbean countries, one of them Cuba. This three-year project began in April 1997.

Proposals for a Cuban national fungal conservation strategy are being produced largely in collaboration with mycologists at the Instituto de Ecología y Sistemática (IES) in Havana, with additional input, particularly from the Jardín Botánico Nacional (HAJB) in Havana. The document draws heavily on the computerised data collected and keyboarded for the Caribbean fungal checklist by these two teams and other Cuban mycologists from the Instituto de Investigaciones Fundamentales en Agricultura Tropicál (INIFAT) and the Instituto de Investigaciones de Sanidad Vegetál (INISAV), both also in Havana. The result is already a bulky and complex document which currently exists only in an unpublished Spanish version (Mena Portales *et al.*, 2000). Until adopted by the appropriate authorities, this is not a strategy, but a series of scientific proposals. This chapter provides a short and annotated English summary of its first part: an assessment of the current state of mycology in Cuba. The following specific topics are examined: threats to fungal biodiversity, current state of knowledge by taxonomic group, habitat and substratum, fungi in protected areas, public awareness of mycology, the lack of environmental legislation for fungi, and resources for mycology. The chapter ends with a few notes on how fungal conservation in Cuba might develop in the next few years.

Threats to fungal biodiversity in Cuba

Most mycologists believe that the best way to conserve fungal biodiversity is *in situ*, through habitat conservation. Cuban mycologists support this general view. In Cuba, most threats to natural habitats are much the same as in many other parts of the world. Deforestation for agriculture, poor farming methods, civil engineering projects such as dams, roads and urban spread, mining and other industrial activities, environmental pollution from fertilisers, pest control, nitrification of water and, simply, rubbish all contribute to the destruction of habitats. Other familiar threats include overexploitation of habitats, and illegal hunting and fishing.

A further factor in recent years has been the rapid growth of tourism in Cuba. This has resulted in increased human pressure on key sites, such as beaches and coral reefs, and picturesque mountain areas. Developments in some of these areas are already causing alarm, since tourism also results in a demand for new hotels, bigger roads, extended airports, and many activities detrimental or potentially detrimental to ecosystems. Even activities billed as ‘ecotourism’ are often not truly sustainable. At the root of all of these threats, arguably, is an ignorance of the true value of biodiversity.

There are, however, additional problems that may particularly affect

Cuban biodiversity. These relate to quarantine and biosecurity, particularly the wrongful introduction of exotic organisms. Apart from casual damage done by ill-informed visitors, there is a controversial political element to these issues. The Cuban government has repeatedly claimed that in addition to waging economic war through the blockade, the USA has introduced various pests and diseases as a form of biological warfare. The most detailed allegations claim that the US government introduced the exotic insect *Thrips palmi*, said to be a vector of plant viruses and an important crop pest in its own right, by aerial spraying over Matanzas province on 21 October 1996. Other allegations concern human dengue fever, swine fever, rusts and smuts of sugar cane, and other exotic diseases which were also said to have been introduced as biological warfare (Castro, 1998). This problem is taken seriously by reputable Cuban scientists, and has been judged sufficiently important to merit inclusion in the Cuban National Study on Biodiversity as a significant threat to the country's biodiversity.

Current knowledge of Cuban fungi by taxonomic group

The Darwin Initiative project's computerisation of fungal specimens in the herbaria of IES and HAJB, together with further information derived from the literature, culture collections and various other sources, has resulted in a database of many thousands of records. At the time of writing, these records are still being edited, but for the Caribbean in general they are likely to exceed 140 000. Of these, perhaps 70 000 will represent Cuban fungi and their associated organisms. Preliminary analysis of the records indicates that over 3870 fungal species are known from Cuba, in nearly 2750 genera, of almost 600 families and 106 orders. These 3870 species, which include lichenised fungi, represent an impressive 5.3% of the species already discovered world-wide, but less than 0.3% of the 1.5 million total number of species estimated by some to exist world-wide (Hawksworth, 1991).

Within these records, all fungal phyla are represented in Cuba with the exception of the Labyrinthulomycota and Hyphochytriomycota. It is possible to compare actual species numbers for each group with world estimates of species numbers for the same groups as provided by Hawksworth *et al.* (1995). These comparisons provide a rough estimate of how well each group known from Cuba has been studied in Cuba. As a percentage of known world numbers, the phyla best represented in Cuba are the Ascomycota (9.4%), Acrasiomycota (8.3%), Zygomycota (7.5%)

Table 16.1. *Orders of fungi that are not represented or poorly represented in records from Cuba*

Not recorded[a]	Poorly recorded
Ceratobasidiales	Cantharellales
Cryptobasidiales	Cortinariales
Cryptomycocolacales	Dacrymycetales
Dimargaritales	Diatrypales
Echinosteliopsidales	Dothideales
Elaphomycetales	Entomophthorales
Lahmiales	Erysiphales
Medeolariales	Exobasidiales
Melanogastrales	Gomphales
Myzocytiopsidales	Graphiolales
Neocallismastigales	Laboulbeniales
Neolectales	Lachnocladiales
Pneumocystidales	Leptomitales
Protomycetales	Lichiniales
Rhipidiales	Phallales
Salilagenidales	Phyllachorales
Schizosaccharomycetales	Platygloeales
Sclerosporales	Rhytismatales
Septobasidiales	Saprolegniales
Spathulosporales	Teloschistales
Taphrinales	Thelephorales
Triblidiales	Trichosphaeriales
	Uredinales

[a]One or two other orders, such as the Cyttariales, with specialised distributions or host ranges not expected to occur in Cuba, have not been listed.

and Dictyosteliomycota (6.5%), and the phyla most poorly represented are the Basidiomycota (2.8%) and Plasmodiophoromycota (2.2%). At class level there is no information about Protosteliomycetes and Trichomycetes, and information about Coelomycetes remains poor. These statistics are, however, to be used with care. For example, within the Basidiomycota there are some relatively well-studied groups, like the Hymenochaetales (15.5%) and Poriales (12.5%), while others are poorly represented, like the Uredinales (0.6%). Orders of fungi that are not represented or poorly represented in records from Cuba are listed in Table 16.1.

Current knowledge of Cuban fungi by habitat and substratum

Purely in terms of numbers of records, plant parts are the best studied fungal substrata in Cuba, and the fungi studied on those plant parts tend to be saprotrophs. Numbers of records can be misleading, however. In a large number of these records, as in other countries, the identification of the plant is inadequate: if the herbarium packet reads, for example, 'on wood' or 'on bark' or 'on dead fallen leaves', with no further information, then all one can say about the associated organism is that it is a plant. By implication, the identification is only to Kingdom level. Even for many large, common and abundant vascular plants, therefore, if the substratum is difficult to identify at the time of collection, the level of knowledge remains generally poor.

There are, furthermore, biases in the records of fungi on plant parts. Most information is about fungi on leaves, twigs, stems, bark and wood, with virtually no information about fungi on roots, buds, flowers, fruits or seeds. Furthermore, most information about fungi on cultivated plants, and other plants of economic importance, such as forest trees, relates to parasitic species: saprotrophic fungi on these plants are almost unstudied. For other forest and scrub plants information about saprotrophic fungi tends to be better, but there is less information about parasitic fungi. Many wayside and ruderal plants, or plants of special ecosystems, are not studied at all. In a country where 50% of the native flora is endemic, the biodiversity implication is clear: the groups of plants which are least studied, and usually totally unstudied, are the endemics and those which are threatened.

A further bias can be detected in the groups of fungi recorded from Cuban plants. Most are Hyphomycetes, or from the families Coriolaceae, Meliolaceae, Hymenochaetaceae and Lentinaceae. That bias extends to known information of fungi on endemic and threatened plants, where most of the little available information relates to Hyphomycetes and the Coriolaceae. This probably only means that mycologists with particular interests have been rather active. The splendid work carried out in recent years on fungi of the royal palm, *Roystonea regia*, shows how many unusual fungi potentially could be found on such endemics (e.g. Mercado Sierra, 1983).

Some mutualistic fungi, such as VA mycorrhizal fungi, are relatively well studied, but there is much less information about others, such as ectomycorrhizal fungi, and virtually nothing is known about endophytic fungi in Cuba. The other great group of mutualistic fungi is those forming lichens. In general, lichens remain inadequately studied in Cuba, particularly those inhabiting rocks, and there is a need for many more collections throughout the country, from different habitats. With the exception of a very few records of species colonising basidiomycetes, fungi parasitic on lichens or on other fungi are almost unrecorded from Cuba.

Given Cuba's excellent recent record in medicine, fungi of clinical importance are well studied, with about 85 species of 36 genera recorded on *Homo sapiens* in the country. Much work is being carried out, not least with opportunistic species associated with AIDS, and particularly at the Pedro Kouri Institute in Havana. Some information is available about fungi allergenic to humans (and other animals), particularly through studies of airborne fungi, but this ecological group needs further study in Cuba, particularly in the interior of the country. Little remains known about many mycoses associated with work conditions, or of environmental origin, or about mycotoxicoses, and Cuban scientists feel it is necessary to intensify techniques for rapid detection and control of opportunistic mycoses.

Knowledge of fungi that develop on other vertebrates is insufficient, and research has so far been directed mainly towards fungi pathogenic on domestic animals and other animals in captivity. Almost nothing is known about fungi associated with but not pathogenic on these animals, or about fungi with any kind of association with wild vertebrates (these include, of course, not only mammals, but birds, fish, reptiles and amphibians). Fungi on dung, similarly, remain practically unstudied in Cuba. There has been some work on fungi associated with insects, particularly bees, but information about fungi on most other insects, and on all other groups of invertebrates is more or less absent.

About 60 species of fungi have been reported from Cuba as growing on non-naturally occurring materials, as agents of biodeterioration or biodegradation. Particular studies include biodegradation of fats, degradation of cellulose, biodeterioration of wood in roofs and of monuments, contamination of microfiches, biodeterioration of museum exhibits (particularly of skulls and other bones), and of books, journals and other documents, and corrosion of plastics, textiles and glass. These investigations are being carried out at a range of centres including the Facultad de Biología Universidad de Havana, the Instituto de Historía de Cuba, the Centro Nacional de Investigaciones Científicas and the Centro de Investigaciones Químicas.

By and large, fungi growing on naturally occurring materials, or in unusual ecosystems, remain almost or completely unstudied in Cuba. About 50 species of phycomycetes were recorded from freshwater in the early 1950s, and there is one work about Ingoldian fungi in foam from Cuban waterfalls. Marine fungi are similarly poorly known, with 43 species recorded in collections of the Instituto de Oceanología in Havana. Nothing is known about fungi inhabiting rocks ('cryptoendolithic fungi') in Cuba. Given their potential importance, soil-inhabiting fungi remain remarkably little studied in Cuba. There are collections of soil fungi, including a few thermophiles, in INIFAT. The only information about fungi from Cuban caves, another interesting but specialised habitat, relates to the human pathogen *Histoplasma capsulatum*.

Fungi in protected areas

Cuba has a national system of nature reserves and protected areas which covers about 22% of the country's area, and all major vegetation types. The level of protection provided, however, varies with different categories of reserve, and in reality, only in about 6% of the country can nature be described as having strict protection. These areas include Natural Reserves, National Parks, Ecological Reserves, certain Natural Monuments, Floristic Reserves, Animal Refuges, and Nature Parks. Although no information is currently available, areas held by the Cuban military, such as training areas or other restricted areas are also likely to be of value in nature conservation. Fungi found in these reserves and other areas, or known to be associated with ecosystems covered by these reserves, are likely to enjoy some sort of protection. Cuba's superb infrastructure, in this sense, however, is a two-edged sword: the tight legislation protecting these ecosystems has meant that getting permission to collect and study fungi in protected areas is often very difficult because of the sheer number of administrative, not to say bureaucratic, hurdles which must be overcome.

Four of the reserves have been accorded 'UNESCO Biosphere' status. In these reserves 950 fungal species, about 25% of all fungi known from Cuba, have been recorded. The biosphere reserve where most fungi have been collected is Cuchillas del Toa, while that with the smallest number of recorded species is Peninsula de Guahanacabibes. In terms of area occupied, the best studied is Sierra del Rosario, followed by Peninsula de Guahanacabibes, Cuchillas del Toa and Baconao, in that order. In biosphere reserves, the best-recorded fungal groups are conidial fungi (250 species), and the families Meliolaceae (63 species), Coriolaceae (55 species) and Hymenochaetaceae (38 species). With some notable exceptions, such as the Alturas de Banao region, information about the status of fungi in other protected areas is almost totally lacking. The lack of material resources in many reserves frequently means that adequate local conservation strategies simply do not exist, even for vascular plants.

Although Cuba's botanic gardens contain no areas set aside specifically for fungi, they do have a role to play in fungal conservation and, when mycologists are present on the garden staff, often quite good information is available about their fungi. Of those which have recorded fungi, the botanic garden with most is HAJB with 171 species, mainly basidiomycetes, and the one with fewest is the Jardín Botanico de Cupeinicú, with 72 species. Some specialist botanic gardens, however, have no information at all about their fungi, but could be interesting and significant. These include the Orchid Garden in Soroa and the Fern Garden in Santiago de Cuba. The fungal group best documented for occurrence in botanical gardens nation-wide is the conidial fungi.

Public awareness

Cuba has a high general level of education. Even so public awareness of the fungi is very low. Partly this is cultural, the Cuban population being markedly different in this respect from those of neighbouring Guatemala and Mexico, with their long traditions of use of fungi in religious rites. In one recent unpublished study of 900 people within a single borough of Havana, significant differences in public perception of the fungi were revealed. The over 40s generally considered them to belong in the plant kingdom, to be a nuisance, and to be associated mainly with plants. The under 40s in comparison tended to understand that they belonged in their own kingdom, could be beneficial, and believed their main significance was medical. The study related these different views directly to different treatments of the fungi in schools and higher education during different periods, suggesting that for mycology education really does make a difference.

In Cuba, at present, the national environmental strategy does not adequately cover the country's fungi. As a result, treatment of the fungi in programmes of basic education is generally insufficient. Only in higher education, where it meets the requirements of different specialisms, is there proper coverage of mycology. Very few of the 174 generalist museums in Cuba provide educational coverage of the fungi. Other organisations combining public recreation and scientific education, such as aquaria, botanic gardens, zoos and cultural centres, rarely or never hold exhibits about fungi. Conservation of fungi gets little direct coverage on radio and television in Cuba, although environmental and nature programmes make occasional references to fungi.

Unlike in Britain, the formation of amateur and other societies is strictly controlled by legislation in Cuba. There is therefore no well-developed network of amateur nature clubs in Cuba, even for bird-watchers, and effective nongovernmental organisations for nature conservation as understood in Britain are also not well developed. Amateur groups for the study of fungi are, *a fortiori*, nonexistent. Scientific societies exist, which to some extent cover mycology. These include the Sociedad Cubana de Botánica, the Sociedad Cubana de Microbiología and the Sociedad Cubana de Fitopatología. These are, however, purely for professional scientists, with no amateur membership, and the need for societies that unite amateurs and professionals is still not fully recognised. Furthermore, although over 140 mycologists are known to be active in Cuba, indicating a clear need, there is still no Cuban Mycological Society.

Environmental legislation

Environmental legislation of one sort or other has existed in Cuba from the pre-revolutionary period, and even from Spanish colonial days. There was considerable legislative activity on this topic during the 1980s, strengthened in the 1990s particularly after the Rio Summit. The various laws, however, while indicating a positive attitude to nature conservation, are rather complex and, at times, conflicting. Cuban environmental experts recognise that their implementation contains many practical problems, and there is concern, for example, that fines are generally derisory. As far as fungi are concerned, current environmental legislation in Cuba provides no specific coverage of their use or conservation, and no specific legal status seems to exist for this group of organisms.

Resources

Cuba's major mycological resource is, of course, its mycologists, but even apart from these, the country has the best infrastructure for work in mycology in the Caribbean and, arguably, the whole of Latin America. There are at least 40 institutions in the country which include at least some research on fungi in their remit (coverage includes systematics, ecology, medical and veterinary mycology, industrial mycology, biotechnology, plant pathology, quarantine, biodeterioration and biodegradation). Some of these institutions have adequate material resources. In addition there is a good network of regional laboratories and extension scientists and technicians.

There are 17 culture collections for fungi in Cuba, including several with specialist interests. The most important is in INIFAT, with about 8000 isolates. This remarkable culture collection, undoubtedly the largest in

Central America, and one of the largest in the tropics, has survived and even grown despite inadequate equipment and frequent power cuts. There are rather more herbaria than culture collections in Cuba. The three main collections are in IES (an estimated 20 000 specimens), HAJB (an estimated 7500 specimens) and INIFAT (an estimated 3000 specimens). Among smaller herbaria, the collections of hyphomycetes in the Centro de Evaluación y Conservación de Ecosistemas Terrestres de Camagüey, and of pathogenic forest basidiomycetes in the Instituto de Investigaciones Forestales in Havana, deserve particular mention.

The United States' blockade, combined with the collapse of the former Soviet Union, has resulted in significant shortfalls in material resources for most if not all of these institutions. There are problems with obsolete scientific equipment (particularly poor-quality microscopes), and difficulties in obtaining reagents and consumables for research. Culture collections lack equipment that, outside Cuba, would be considered essential: freeze-drying techniques are unavailable, and cultures have to be maintained in oil, or regularly subcultured on agar, both highly labour-intensive techniques. Chemicals for fumigation of herbaria cannot be obtained from the cheapest source (the USA) and are therefore more expensive, more difficult, and sometimes even impossible to obtain. The result has been the loss, over the last twenty or so years, of significant numbers of type specimens to the depredations of insects. Expedition vehicles tend to be old fuel-inefficient Soviet-built jeeps and trucks for which spare parts are now largely unavailable. These, combined with a general shortage of fuel, make access to distant places and places rich in biodiversity very difficult. For the present Darwin Initiative project, much of the day-to-day travel between institutes in and around Havana has used an ancient Russian-built 'Jupiter' motorbike and sidecar – pleasant enough travelling in the tropics, but disconcerting when being overtaken by energetic cyclists.

There are also severe informational problems. Institutional libraries have great problems in maintaining up-to-date runs of even major scientific journals. Essential literature, for example catalogues like the *Bibliography of Systematic Mycology*, and the *Index of Fungi*, are incomplete, even in the country's main systematic institutions. Specialist literature on particular taxonomic groups is also generally lacking. The scientific visitor to Cuba soon encounters the resulting thirst for new literature, and rapidly learns to pack his or her bag with books on the outward journey, usually coming home with piles of mail which Cuban colleagues are unwilling to entrust to their own postal services. Cuban scientists face additional problems in publishing their work. Until very recently, one distinguished Cuban mycologist wrote all his papers using a typewriter made in 1909. When a manuscript is successfully sent to a journal, perhaps by hand-delivered mail through a returning foreign visitor, all too frequently an editor may return the manuscript for corrections, entrusting it to conventional post without imagining the delays, difficulties, frustrations and dismay this can cause. One superb and major mycological manuscript by Cuban and Czech authors waited nine years before it could be published in 1998.

Computers and computing resources remain another very important limiting factor. With the exception of certain flagship reserves, computers are almost entirely lacking from most nature reserve offices, making it difficult for staff there to organise their reserve records efficiently, or to produce machine-readable reserve management plans. Access to computers is also severely limited in many research institutions, and enormous frustrations can result from loss of computing time because of power cuts and other shortages. Another computing and informational problem is e-mail and Internet access. Various difficulties continue to hinder easy access for researchers and technicians to these services.

Some possible future developments

This chapter has briefly reviewed the current state of mycology and fungal conservation in Cuba. Although it would be premature to comment on the proposed strategy itself, a few remarks about future directions may be appropriate. Fungal conservation is only likely to happen if there are people who care enough to make it happen. Those people are mycologists. In many parts of the world mycologists are an endangered species. It follows that fungal conservation can only occur if mycologists are conserved. That means that Cuba with its wealth of expertise is likely to become increasingly important. There is a very good case for directing resources to its mycologists. This is something the British, Mexicans and Spanish in particular are already doing, but not in sufficient quantities. Cuba, nevertheless, can do much to help itself here, through the formation of a Cuban Mycological Society. This single act would, at present, do more than any other to forward fungal conservation in that country, and the Cuban authorities are urged to facilitate establishment of that society.

Furthermore, assistance to Cuban mycologists is a matter that should transcend the political rancour which has existed between the USA and Cuba. The restrictions imposed on US scientists by their own administration through the blockade are having a deleterious effect on their own important scientific projects. There is, for example, a large US-funded project on the flora of the Greater Antilles (and this flora includes fungi and lichens). But it is almost completely impossible for the organisers of this project to access and include recent information about the biggest and, in biological terms, most important island in the archipelago. The efforts being made by some scientists in the USA to repatriate botanical data to Cuba (<http://www.ascoll.org/cuban3.html>) are to be applauded. It is a pity that this has not yet extended to fungal records.

Although much has been done for Cuban mycology over the last three years, an enormous amount remains to be done. Computerising the floristic records and producing a checklist is only a start. Subsequent stages should include the production of electronic distribution maps of fungi, Red Data lists of species known to be endangered or of limited distribution, grey lists setting out the plants, plant parts and other substrata for which few or no fungi have been recorded, and many other uses of that information. Further in the future, but not too distant, should be the large-scale production of electronic keys, descriptions and expert systems, all fully illustrated with digitised photographs. There is furthermore a clear need for a simple guide to the identification of dead fallen leaves by shape and size, and of dead fallen twigs and wood by anatomy observable with a hand-lens and other field characteristics. Such a guide would surely improve the proportion of future records with an adequate identification of the associated plant. Even if a guide like that were available, however, field identification of dead parts of grasses, sedges and similar plants would still usually be impractical.

The growth of tourism in Cuba also provides an urgent stimulus to survey and assess particular areas rich in biodiversity and of outstanding natural beauty. British mycologists are justifiably proud of a small number of sites where fungi have been intensively recorded, such as Esher Common or Slapton Ley. No equivalent sites exist anywhere in the tropics. Given the splendid human resources Cuba has in terms of high calibre systematists, not only in mycology, but also in many other areas of biology, a good case can be made for combining such tourism impact surveys with the establishment of a small number of such sites in the tropics, at least for the fungi, perhaps also for one or two other groups. This would, of course, fall far short of an all taxon biodiversity inventory, but it could mark the first steps in that direction. Localities that spring to mind as suitable include the mogotes hills near Viñales, the swamps of Ciénaga de Zapata, and the reserve of Alturas de Banao, all mentioned earlier in this chapter. There are, doubtless, many other appropriate sites.

Conclusion

Any scientist visiting Cuba is likely to come away with a whole spectrum of strong impressions. Among these, the high quality of Cuban science, and the calibre, enthusiasm and dedication of its scientists who achieve wonders with minimal resources, and the well-developed infrastructure are likely to be prominent. The *Cuban National Study on Biodiversity*, produced before this Darwin Initiative project began, and without any external assistance, is a profoundly impressive work. It is hard to imagine any other country with so low a per capita GNP being able to produce such a document unaided. The apparent ease with which Cuban colleagues were able to pull together huge amounts of information for the much more detailed proposed Cuban national conservation strategy for fungi was, frankly, amazing. It must be evident that, while Cuban mycologists need material resources, they do not need aid. They are seeking collaboration from a strong position, and have much to contribute. It means that, in Cuba, perhaps more than in any other tropical country in the world, there is a real chance to carry out real fungal conservation over the next few years.

Acknowledgements

The UK Darwin Initiative is warmly thanked for support through its project Fungi of the Caribbean. Part of this chapter constitutes an English language summary of parts of the proposed Cuban national conservation strategy for fungi (Mena Portales *et al.*, 2000), and is thus derived from the work, thoughts and ideas of my Cuban colleagues and friends. Because I have annotated that summary with my own comments and interpretations (including some with which they may not agree), it has seemed better to offer the present chapter under my sole authorship. It is therefore only appropriate to acknowledge with thanks their contribution to this work.

References

Arnold, G. R. W. (1986). *Lista de Hongos Fitopatógenos de Cuba*. 206 pp. Edit. Científico Técnica: Ciudad de La Habana, Cuba.

Castro, F. (1998). Informe central al V Congreso del Partido Comunista de

Cuba. *Granma Internacional* **32** (47), 3–14.
Hawksworth, D. L. (1991). The fungal dimension of biodiversity: magnitude, significance, and conservation. *Mycological Research* **95**, 641–655.
Hawksworth, D. L., Kirk, P. M., Sutton, B. C. & Pegler, D. N. (1995). *Ainsworth & Bisby's Dictionary of the Fungi*. Eighth edn, 616 pp. CAB International: Wallingford, Oxon.
Kreisel, H. (1971*a*). Clave para la identificación de los macromicetos de Cuba. *Ciencias* Serie Ciencias Biológicas **4** (16), 1–101.
Kreisel, H. (1971*b*). Clave y catálogo de los hongos fitopatógenos de Cuba. *Ciencias* Serie Ciencias Biológicas **4** (20), 1–104.
Mena Portales, J., Herrera Figueroa, S., Mercado Sierra, Á. & Minter, D. W. (2000). *Estrategia para la Conservación de la Diversidad Fúngica en Cuba. Estado de Conocimiento, Estrategia y Plan de Acción*. Unpublished report, 154 pp. Instituto de Ecología y Sistemática: La Habana, Cuba and CABI Bioscience: Egham, Surrey, UK.
Mercado Sierra, Á. (1983). La palma real (*Roystonea regia*): un sustrato idóneo para el desarrollo de hifomicetes demaciáceos. *Acta Botánica Cubana* **15**, 1–13.

17

Microfungus diversity and the conservation agenda in Kenya

P. F. CANNON, R. K. MIBEY & G. M. SIBOE

Introduction

Microfungi are rarely considered within conservation policies for a number of reasons: they are small, poorly known (especially in the tropics), extremely diverse, and their fruit bodies are often ephemeral. They are frequently perceived at best as not charismatic, and at worst as threats to other species. However, it is probably true to say that the majority of fungi would be describable as 'microfungi' and many are likely to be rare and threatened. Nevertheless, they may play important roles in the ecosystem through positive interactions with other organisms, and they represent an enormous range of genetic and metabolic resources. Thus, consideration of microfungi in relation to the issues that are addressed in conservation programmes for African animals and plants is appropriate, and represents a valuable model for many other organism groups.

One of the principal barriers to the inclusion of microfungi as targets for conservation is knowing whether species are genuinely rare, or simply rarely recorded. With the assistance of the UK Government's Darwin Initiative, we have addressed this problem in Kenya by collecting and studying fungal species which are associated with rare and endangered plants, and which are likely to be host limited. We can then be confident that the fungi are at least narrowly distributed, and are threatened to at least the same degree as their plant hosts. We have identified a range of biotrophic species associated with rare Kenyan plants, many of which are being published as new taxa, and are now monitoring populations and establishing host ranges by searching for them on related host plants. We hope then to be in a position to introduce microfungi to the conservation agenda, by demonstrating that the loss of particular plant species from the Kenyan flora will result in the loss also of associated fungi. Plants are actually community microcosms on which many other organisms depend. We hope that this realisation will lead to enhanced consideration by conservation decision-makers in Kenya of the small-bodied but species-rich groups such as fungi and insects.

The fungi comprise one of the most species-rich and ecologically important organism groups on Earth. Despite this, until recently they have rarely been considered seriously in the course of *in situ* conservation decision-making, especially in tropical regions, and funding for *ex situ* conservation in the form of culture collections remains inadequate (Kirsop & Canhos, 1998). The reasons for this are not difficult to discover. Firstly, the conservation movement, largely funded through voluntary donations, has historically concentrated upon large charismatic organisms with which the general public empathises. Popular perceptions of fungi tend to revolve around toxicity and disease, with few species considered in a positive light. Secondly, there are significant problems in analysis and monitoring of fungal populations, and in the sheer number of potential candidate species for conservation. Nevertheless, there are compelling reasons to include fungi in conservation strategy, and the project outlined here aims to provide basic data to overcome one major obstacle, that of deciding whether fungi which are rarely recorded are genuinely rare or simply rarely detected.

Why are fungi difficult to conserve?

There are very large numbers of fungal species, making prioritisation complex and time-consuming. Estimates have been the subject of vigorous debate over the last ten years, but most agree that at least 1 million and perhaps as many as 1.5 million species exist world-wide (Hawksworth, 1991; Rossman, 1994). Predictions of the number of species occurring in particular localities or habitats are also difficult as fungal distribution patterns do not necessarily mirror those of better known organism groups. However, two protected sites in southern Britain, Esher Common and Slapton Ley National Nature Reserve, both harbour between 2500 and 3000 recorded species with only about 40% overlap (D. L. Hawksworth, J. A. Cooper, pers. comm.), suggesting that substantial numbers remain to be discovered at both sites. Even at the individual sample level, species numbers can be overwhelming, with 50–100 commonly detected from individual substratum collections, even where the techniques used do not result in a complete survey (e.g. Bills & Polishook, 1994; Lodge & Cantrell, 1995). There is very little reliable information which can be used to contrast fungal diversity in tropical and temperate regions (Lodge *et al.*, 1995), but as a large proportion of species is host specific to some degree, it is likely that fungal diversity shows a similar bias towards the tropics as does the diversity of many other, better studied organisms (Hawksworth & Kalin-Arroyo, 1995).

Fungi are also poorly known. Only an estimated 5% have received even a rudimentary description (Hawksworth, 1991; Cannon & Hawksworth, 1995). Even of these, the vast majority has not been treated in a modern monograph, and information that is available is inaccurate, sketchy and often buried in inaccessible journals. Fungal classification has undergone a series of major changes in recent years with the incorporation of lichens into mainstream fungal systems, the continuing integration of anamorph and teleomorph classifications, and now the molecular revolution which is providing dramatic new evidence of relationships. These are all crucially important advances, but in the short term the instability makes life even more difficult for conservationists and other applied biologists. Species concepts are poorly understood, with traditional morphology-based systems being replaced by nucleic acid-based phylogenies and biological species concepts (Brasier, 1997). This has particular relevance for necrotrophic plant pathogens, with great uncertainty over the extent of host specificity in many genera (Rehner & Uecker, 1994; Cannon *et al.*, 2000).

Almost all fungi exist primarily as mycelium, often immersed within host tissue, with fruit body production (currently essential for both detection and identification) ephemeral and greatly influenced by external factors. This means that it is often difficult to establish whether a given species is genuinely absent from a location, or is simply behaving cryptically. Both fruit bodies and colonies are extremely small in many cases, and most are not conspicuously pigmented. Detection and subsequent monitoring is therefore sometimes a painstaking process, involving careful microscopic observation over an extended period.

Why are fungi important?

Fungi and bacteria constitute the dominant organism groups involved in C/N cycling, primarily in the degradation of organic matter (Christensen, 1989). Critical stages in lignin decomposition are almost exclusively carried out by fungi, and the breakdown of complex biomolecules such as cellulose and tannins in soils is due mostly to fungal enzymatic activity (Cooke & Whipps, 1993). Hyphal length has frequently been estimated at over 1 km g^{-1} in grassland soil (Kjøller & Struwe, 1982), and fungi account for up to 90% of total living biomass in forest soils (Frankland, 1982).

Fungi mediate plant and animal health and growth via symbiotic associations. These may be mutualistic, for example mycorrhizas, endophytes and ant-nest symbionts (Read *et al.*, 1992; Redlin & Carris, 1996; Chapela *et al.*, 1994), or antagonistic with a plethora of well-known plant and animal diseases. They provide a food source for a wide range of vertebrates and invertebrates including insects, mites, molluscs and nematodes (Shaw, 1992), and gain nutrition in turn from living or dead animals as well as plants.

Large numbers of fungal species, especially saprotrophs and necrotrophic taxa, can be isolated easily into pure culture. This has led to extensive screening for chemicals with useful pharmaceutical and industrial properties. Some of the most valuable and widely used pharmaceuticals from natural products originate from fungi, including the penicillin antibiotics, the antifungal cyclosporins used in transplant surgery, and recent 'cholesterol-busting' drugs based on zaragozic acids (Dreyfuss & Chapela, 1994; Bergstrom *et al.*, 1995).

How can we include fungi in conservation strategies?

Most fungal species are poorly known, if they are described at all. Even when species are known, the difficulties of sampling and recognition mean that it is problematical to assess whether species that have been detected only a small number of times are really rare, or merely under-recorded.

In Kenya, we have been addressing this issue with the assistance of a major grant from the UK Department of the Environment's Darwin Initiative. The approach taken is to target fungi which are likely to be host specific, and which are associated with plant species which are themselves rare or endangered. We can then be confident that these fungal species are at least as rare as their plant partners are. The aim is to provide information that can be used by policy-makers in prioritising conservation activities; the project leaders have no direct influence on national and regional environmental protection strategies. However, we are in a position to tell those who do make the decisions that if particular plant species become extinct in Kenya, fungal species will also be lost.

The information may also be used to prioritise protection for particular populations of plants. The distribution of fungi associated with endangered plants may not be coincident with that of their hosts; some populations may lack the associated fungi. In Kenya as in many tropical countries pressure on native vegetation is intense due to a rapidly increasing population with burgeoning needs for farmland, timber for construction, firewood etc. Hard decisions sometimes have to be taken to sacrifice stands of natural vegetation to meet the needs of the people at large. In those circumstances, decisions may be taken to promote conservation of rare plant populations which harbour endangered fungal taxa.

Most host-specific species of fungi are biotrophic or necrotrophic, gaining their nutrition from living plant tissue. The relationship with the plant is by no means always deleterious. The fungus may compensate the plant for providing nutrients and shelter by secreting toxic metabolites that discourage herbivory by vertebrate or invertebrate animals; this is particularly well documented for endophytic fungi associated with grasses (Redlin & Carris, 1996), but may be a widespread phenomenon. It is rare in natural environments for an indigenous fungus to have a serious deleterious effect on a native plant species (Allen *et al.*, 1999); good parasites don't kill their hosts. Where significant damage does occur, it is often through the introduction of alien species, or modifications of the environment that upset the balance between fungus and plant. Positive contributions to ecosystem function by parasitic fungi are increasingly being considered by plant pathologists (Browning, 1974; Helfer, 1993; Ingram, 1999), leading to calls for their conservation. Coevolution of pathogens with wild crop hosts has resulted in a wide range of disease resistance mechanisms, which have not always been utilised effectively in crop breeding (Wood & Lenné, 1997; Allen *et al.*, 1999). Wild relatives of crop plants are therefore acquiring new significance in world economics, and genetic manipulation systems already allow the transfer of pathogen resistance between unrelated plants, perhaps providing new roles for genes from native plants in food production.

The need for integrated conservation action plans

Species cannot be preserved *in situ* in isolation. If rare or endangered plants are to be protected, an array of associated species must also be taken into account. A wide variety of organisms may be dependent for survival on preservation of the plant species. These may include the following.

Biotrophic and necrotrophic fungi, bacteria, other plants etc. directly associated with the plant.

Organisms using the plant as a platform, e.g. foliar epiphytes and lichenised fungi on bark.

Endophytes, e.g. fungi, bacteria and viruses within plant tissues.
Root-inhabiting nematodes, protozoans, etc.
Bark and stem-boring insects, leaf-cutting ants and wasps, leaf-miners, gall-formers, etc.
Pollinators (insects, birds, bats, etc.).
Yeasts and other organisms associated with nectar.
Fruit-eating/ingesting organisms, from mammals to bacteria.
Organisms eating, degrading or taking shelter in dead plant material.

In some circumstances, for example mycorrhizas or pollinators, the loss of the associated organism will cause death or at least have a serious deleterious effect on survival of the plant. In other cases, the dependence is only one-way, but loss of the plant will result in multiple extinctions.

Health audits of endangered plants

Many organisms do not have a beneficial effect on plants. In most cases their prominence in natural ecosystems may be due to human intervention in the past or present (Allen *et al.*, 1999), such as encroaching agriculture, logging or firewood collection, or industrial pollution. Major environmental modifications such as those associated with global warming are likely to result in more extreme perturbations of the natural balance between plants and other organisms.

It is valuable to perform health audits on rare or endangered plant species, identifying species alien to their community such as plant pathogenic fungi and bacteria spreading from neighbouring agricultural areas, immigrating root-, stem- and bark-boring insects, or invading weeds. It is not always easy to tackle the sources of these threats to plant health and existence (and indirectly to the conservation of all dependent organisms), but their identification may enable protective measures to be taken. It is also difficult in some circumstances to decide whether fungi are genuinely native to the region or have been introduced: the distinctions between indigenous and alien populations may be evident only at the genetic level. This is particularly the case with necrotrophic fungi such as *Colletotrichum*, *Guignardia* and *Phomopsis*, all of which are commonly encountered in surveys of natural habitats.

Darwin Project activities

The first step was to establish an initial list of target plants. Plant conservation research in Kenya has been centred on the National Museums in

Nairobi, and several documents are available which detail threatened taxa. Prominent amongst these are accounts by Gillett (1979) and Luke (1991), and much of the information was usefully summarised by Beentje (1994), who included IUCN threat categories in his treatment of the woody plants of Kenya, and also by Walter & Gillett (1997). Available fascicles of the *Flora of East Tropical Africa* were also surveyed for endemic and/or threatened plants. A second category of plants was also included in response to requests from the Kenyan Forestry Department, species restricted to the few remaining remnants of Guinea–Congolian rainforest in western Kenya, centred on Kakamega Forest. This vegetation has links with West rather than East Africa. While many of the species in the Kakamega region have widespread distributions, their presence in Kenya is threatened and extensive fragmentation of the forested regions west of Lake Victoria means that populations are genetically isolated. Forests in the Kakamega region are seriously threatened due to pressure from the rapidly increasing surrounding population, and have been the subject of planning for sustainable exploitation in the past (Wass, 1995). Plants from the coastal Kaya forests of Kenya were also particularly considered, to build on previous work by the National Museums of Kenya and the World Wide Fund for Nature (Robertson & Luke, 1995). Kayas are forests or woodland fragments of religious significance to local inhabitants, which have historically been well preserved but are now in some instances being threatened as traditional values are eroded. Finally, a small list of charismatic plant species was included, which, while not necessarily endangered, are well known to many nonspecialists. This included the arborescent *Lobelia* and *Senecio* species of Mount Kenya and other highland regions, and also species of *Protea* with their spectacular flowers. Target species were prioritised using the following criteria: conservation status (e.g. restricted distribution, known threats); woody habit (provides more microhabitats for fungal colonisation); systematic isolation (plants more likely to harbour host-specific fungi); ease of recognition in the field; distinctiveness of habitat; ease of access.

The research programme has surveyed forests in a number of regions of Kenya, including Kakamega Forest, the Cherangani Hills, Mount Elgon, the Aberdares, Mount Kenya, the Taita and Shimba Hills, and Arabuko-Sokoke Forest. More information can be found in the project website, <http://www.cabi.org/bioscience/darwin.htm>. Approximately 110 plant species have been surveyed, and each harboured at least one identifiable fungus, with many of them being host to many species of fungi. Surveys concentrated on living leaves to detect biotrophic and necrotrophic *Ascomycota*, including the mitotic forms traditionally referred to as hyphomycetes and coelomycetes. Some studies were also made of fallen leaves, many fungi from which would previously have been present in living tissues as endophytes, and of dead twigs attached to living plants, often dying back through fungal activity.

The fungi of particular note belong to the following assemblages:

Meliolaceae (*Ascomycota*: *Meliolales*). The sooty moulds have been the subject of a previous major study of fungi from natural habitats in the Shimba Hills (Mibey & Hawksworth, 1997), and it is notable that a number of further new taxa have been identified from the same region during the current project, even after the intensive sampling which preceded the first report. Twelve new species have been described as part of the current project (Mibey & Kokwaro, 1999; Mibey & Cannon, 1999), eleven of which belong to the genus *Meliola* and one to *Cryptomeliola* (the latter on the same plant as one of the *Meliola* species). There is some controversy surrounding the host specificity of the *Meliolaceae*, but it seems likely that coevolution has resulted in at least a degree of specialisation (Mibey & Cannon, 1999).

Phyllachoraceae (*Ascomycota*: *Phyllachorales*). This family has not been studied in detail from Kenya, but a survey during the current project has so far identified 40 taxa, including six new species (Cannon, in prep.). The new genus *Fremitomyces* has also been described from Kenyan material (Cannon & Evans, 1999). Most species of the family are biotrophic, and thus generally considered to be host specific at least at the genus level. In addition, colonies of *Colletotrichum* species (sometimes with their *Glomerella* teleomorphs) have been found associated with lesions on many of the rare indigenous plants surveyed. Most are referable to the species aggregate *C. gloeosporioides*, which is extremely widespread and a common disease-causing organism of crop plants. However, there are well-established genetic clusters within the aggregate as a whole, some of which are clearly host specific (Cannon *et al.*, 2000). Regrettably, the project resources did not allow culture and molecular characterisation of the strains from indigenous plants, so we are unsure whether they constitute part of the natural environment or have invaded from elsewhere.

Asterinaceae (*Ascomycota: Dothideales*). Also monographed by Mibey & Hawksworth (1997), we have amassed around 40 further collections, again including further taxa from the Shimba Hills as well as many from the Kakamega region. A number include only the anamorph, a part of the life cycle which has not been studied intensively, so species concepts are difficult to establish.

Cercospora and relatives (*Ascomycota: Dothideales*). Around 20 species have been recognised as undescribed, with several already prepared for publication (Siboe, Kirk & Cannon, 1999; Siboe *et al.*, 2000). Most are necrotrophic pathogens found associated with leaf spots.

A very wide range of other fungi has also been studied. Some of the better known groups of biotrophic fungi such as the *Erysiphales* (powdery mildews) are not typical of forested habitats where most of the research has been carried out. Rusts have been rarely recorded for the same reason, though one species of *Aecidium* from *Diospyros greenwayi* and *D. shimbaensis* from the Shimba Hills is of particular note as it appears not to parasitise *D. kabuyeana*. The first of these plant species is in the IUCN 'rare' conservation category, and the other two are classed as 'vulnerable' (Beentje, 1994). The fungus is similar but not identical to an Indian species on other species of *Diospyros*. Research continues.

Conclusions

This is a distinctive research programme, which is breaking new ground in some respects. So far as we are aware, fungi have not previously been integrated into plant conservation strategies in this way before, either in the tropics or in temperate regions. While we do not pretend to be able to provide unequivocal answers during the programme, we will be able to provide strong pointers that will facilitate the inclusion of fungal species in protection strategies, and explore the complex relationships between fungi and plants.

The programme charts a middle course between conservation of individual species and protection of environments. The first approach has many drawbacks due to the complex interrelationships between target species and other organisms, and the second provides little information about individual species.

The project represents a small beginning, in a research area that is potentially far-reaching. If this approach proves fruitful there is considerable potential to expand the programme to cover other plant species in Kenya and beyond. It would also be most valuable to integrate these studies with parallel research into other species-rich organism groups associated with rare plants, particularly the insects. This would result in developing knowledge of the complex food webs that underpin biodiversity, and gain much information about the component parts of threatened ecosystems.

References

Allen, D. J., Lenné, J. M. & Waller, J. M. (1999). Pathogen biodiversity: its nature, characterisation and consequences. In *Agrobiodiversity: Characterisation, Utilisation and Management* (ed. D. Wood & J. M. Lenné), pp. 123–153. CAB International: Wallingford, UK.

Beentje, H. (1994). *Kenya Trees, Shrubs and Lianas*. 722 pp. National Museums of Kenya: Nairobi, Kenya.

Bergstrom, J. D., Dufresne, C., Bills, G. F., Nallin-Omstead, M. & Byrne, K. (1995). Discovery, biosynthesis and mechanism of action of the zaragozic acids: potent inhibitors of squalene synthase. *Annual Review of Microbiology* **49**, 607–639.

Bills, G. F. & Polishook, J. D. (1994). Abundance and diversity of microfungi in leaf litter of a lowland rain forest in Costa Rica. *Mycologia* **86**, 187–198.

Brasier, C. M. (1997). Fungal species in practice: identifying species units in fungi. In *Species: the Units of Biodiversity* (ed. M. F. Claridge, H. A. Dawah & M. R. Wilson), pp. 135–170. Chapman & Hall: London.

Browning, J. A. (1974). Relevance of knowledge about natural ecosystems to development of pest management programs for agro-ecosystems. *Proceedings of the American Phytopathological Society* **1**, 191–199.

Cannon, P. F., Bridge, P. D. & Monte, E. (2000). Linking the past, present and future of *Colletotrichum* systematics. In *Colletotrichum: Host Specificity, Pathology and Host-Pathogen Interaction* (ed. D. Prusky, S. Freeman & M. B. Dickman), pp. 1–2. APS Press: St Paul, Minnesota.

Cannon, P. F. & Evans, H. C. (1999). Biotrophic species of *Phyllachoraceae* associated with the angiosperm family *Erythroxylaceae*. *Mycological Research* **103**, 577–590.

Cannon, P. F. & Hawksworth, D. L. (1995). The diversity of fungi associated with vascular plants: the known, the unknown and the need to bridge the knowledge gap. *Advances in Plant Pathology* **11**, 277–302.

Chapela, I. H., Rehner, S. A., Schultz, T. R. & Mueller, U. G. (1994). Evolutionary history of the symbiosis between fungus-growing ants and their fungi. *Science* **266**. 1691–1694.

Christensen, M. (1989). A view of fungal ecology. *Mycologia* **81**, 1–19.

Cooke, R. C. & Whipps, J. M. (1993). *Ecophysiology of Fungi*. Blackwell: Oxford.

Dreyfuss, M. M. & Chapela, I. H. (1994). Potential of fungi in the discovery of novel, low-molecular weight pharmaceuticals. In *Discovery of Natural Products with Therapeutic Potential* (ed. V. P. Gullo), pp. 49–80. Butterworth-Heinemann: Boston.

Frankland, J. C. (1982). Biomass and nutrient recycling by decomposer basidiomycetes. In *Decomposer Basidiomycetes* (ed. J. C. Frankland, J. N. Hedger & M. J. Swift), pp. 241–261. Cambridge University Press: Cambridge, UK.

Gillett, J. B. (1979). Kenya. Part of appendix to: Possibilities and needs for conservation of plant species and vegetation in Africa. In *Systematic Botany, Plant Utilization and Biosphere Conservation* (ed. I. Hedberg), pp. 93–94. Almqvist & Wicksell International: Stockholm.

Hawksworth, D. L. (1991). The fungal component of biodiversity: magnitude, significance and conservation. *Mycological Research* **95**, 641–655.

Hawksworth, D. L. & Kalin-Arroyo, J. (1995). Magnitude and distribution of biodiversity. In *Global Biodiversity Assessment* (ed. V. H. Heywood & R. T. Watson), pp. 107–191. Cambridge University Press: Cambridge, UK.

Helfer, S. (1993). Rust fungi – a conservationist's dilemma. In *Fungi of Europe: Investigation, Recording and Conservation* (ed. D. N. Pegler, L. Boddy, B. Ing & P. M. Kirk), pp. 287–294. Royal Botanic Gardens: Kew, UK.

Ingram, D. S. (1999). Biodiversity, plant pathogens and conservation. *Plant Pathology* **48**, 433–442.

Kirsop, B. & Canhos, V. P. (1998). The economic value of microbial genetic resources. Proceedings of a World Federation for Culture Collections workshop, Halifax, Canada, Aug. 12 1998. pp. 1–33. World Federation for Culture Collections: Halifax, Canada.

Kjøller, A. & Struwe, S. (1982). Microfungi in ecosystems: fungal occurrence and activity in litter and soil. *Oikos* **39**, 391–422.

Lodge, D. J. & Cantrell, S. (1995). Fungal communities in wet tropical forests: variation in time and space. *Canadian Journal of Botany* (supplement) **73**, S1391–S1398.

Lodge, D. J., Chapela, I., Samuels, G. J., Uecker, F. A., Desjardin, D., Horak, E., Miller, O. K., Hennebert, G. L., DeCock, C. A., Ammirati, J., Burdsall, H. H., Kirk, P. M., Minter, D. W., Halling, R., Laessoe, T., Mueller, G., Oberwinkler, F., Pegler, D. N., Spooner, B., Petersen, R. H., Rogers, J. D., Ryvarden, L., Watling, R., Turnbull, E. & Whalley, A. J. S. (1995). A survey of patterns in fungal diversity. In *Lichens – A Strategy for Conservation*.(ed. C. Scheidegger & P. Wolseley), *Mitteilungen der Eidgenossischen Forschunganstalt für Wald-, Schnee- und Landschaft* **70**, 157–173.

Luke, W. R. Q. (1991). A preliminary list of rare, vulnerable and endemic plants for Kenya (appendix B cntd.). In *The Costs, Benefits and Unmet Needs of Biological Diversity Conservation in Kenya*, p. 25. Overseas Development Administration.

Mibey, R. K. & Cannon, P. F. (1999). Biotrophic fungi from Kenya: ten new species and some new records of *Meliolaceae*. *Cryptogamie, Mycologie* **20**, 249–282.

Mibey, R. K. & Hawksworth, D. L. (1997). *Meliolaceae* and *Asterinaceae* of the Shimba Hills, Kenya. *Mycological Papers* **174**, 1–108.

Mibey, R. K. & Kokwaro, J. O. (1999). Two new species of *Meliola* (Ascomycetes) from Kenya. *Fungal Diversity* **2**, 153–157.

Read, D. J., Lewis, D. H., Fitter, A. & Alexander, I. (1992). *Mycorrhizas in Ecosystems*. CAB International: Wallingford, UK.

Redlin, S. C. & Carris, L. M. (1996). *Endophytic Fungi in Grasses and Woody Plants*. APS Press: St Paul, Minnesota.

Rehner, S. A. & Uecker, F. A. (1994). Nuclear ribosomal internal transcribed spacer phylogeny and host diversity in the coelomycete *Phomopsis*. *Canadian Journal of Botany* **72**, 1666–1674.

Robertson, S. A. & Luke, W. R. Q. (1995). The coast forest survey of the national museums of Kenya and the World Wide Fund for Nature. In *Conservation of Biodiversity in Africa. Local Initiatives and Institutional Roles* (ed. L. A. Bennun, R. A. Aman & S. A. Crafter), p. 89. National Museums of Kenya: Nairobi.

Rossman, A. Y. (1994). A strategy for an all-taxa inventory of fungal biodiversity. In *Biodiversity and Terrestrial Ecosystems* (ed. C.-I. Peng & C. H. Chou), pp. 1691–1694. Academia Sinica: Taipei.

Shaw, P. J. A. (1992). Fungi, fungivores and food webs. In *The Fungal Community. Its Organization and Role in the Ecosystem* (ed. G. C. Carroll & D. T. Wicklow), pp. 295–310. Marcel Dekker: New York.

Siboe, G. M., Kirk, P. M. & Cannon, P. F. (1999). New dematiaceous hyphomycetes from Kenyan rare plants. *Mycotaxon* **73**, 283–302.

Siboe, G. M., Kirk, P. M., David, J. C. & Cannon, P. F. (2000). Necrotrophic fungi from Kenyan endemic and rare plants. *Sydowia* **52**, 286–304.

Walter, K. S. & Gillett, H. J. (1997). *IUCN Red List of Threatened Plants*. IUCN: Cambridge, UK.

Wass, P. (1995). The Kenyan indigenous forest conservation project: an integrated approach to site conservation. In *Conservation of Biodiversity in Africa. Local Initiatives and Institutional Roles* (ed. L. A. Bennun, R. A. Aman & S. A. Crafter), pp. 271–285. National Museums of Kenya: Nairobi.

Wood, D. & Lenné, J. M. (1997). The conservation of agrobiodiversity on-farm: questioning the emerging paradigm. *Biodiversity and Conservation* **6**, 109–129.

18

Fungi and the UK Biodiversity Action Plan: the process explained

L. V. FLEMING

Introduction

Lead partners, champions, priority species, SAPs, HAPs, and Country Groups are just some of a bewildering array of terms, jargon and acronyms that confront a novice to the UK Biodiversity Action Plan (BAP) (see Table 18.1). This chapter aims to give the background to the BAP, to explain in simple terms what is actually happening (focusing on the production of species and habitat action plans), and will consider the implications of this process for the conservation of fungi in the United Kingdom (UK).

Background

The United Nations Conference on the Environment and Development (the 'Earth Summit') in Rio de Janeiro, 1992, resulted in the widely publicised Convention on Biological Diversity, to which the UK is a signatory. One of the many requirements of the Convention was that individual states produce their own action plans or programmes for the conservation of biodiversity. Accordingly, in 1994, the UK Government published *Biodiversity: the UK Action Plan* (Anon., 1994). This document provided a summary of the state of UK biodiversity and the actions being taken to conserve it. Most importantly, it contained in a final section a set of '59 steps' to be implemented by Government and its agencies. These 59 steps encompassed a range of activities, from the need to continue with the programme of site designations required by European Community Directives, to education and awareness programmes, to actions for UK overseas territories and for the limitation of UK fishing activities. Whilst these actions were a mixture of the ongoing and the new, one step in particular was to lead to a flurry of new activity. This step (33) reads:

Table 18.1. *Glossary of terms used in the Biodiversity Action Plan*

Agenda 21	A programme for action for the 21st century, arising from the Rio conference, to have sustainable development at its core, and to be taken forward by Local Governments, amongst others, in consultation with their citizens.
BIG	Biodiversity Information Group.
Champion	Any organisation or individual who wishes to support the work of an action plan by providing resources, in cash or in kind.
Contact point	Government agency, responsible for acting as initial contact for an action plan, for stimulating action and maintaining standards with regard to the requirements of a published plan.
Country Group	The biodiversity steering group for each country in the UK.
HAP	Habitat Action Plan.
Key habitats	The name, now superseded, for those habitats for which action plans will be/have been prepared.
LBAP	Local Biodiversity Action Plan.
Lead partner	The organisation responsible for leading and co-ordinating the implementation of an action plan.
Long list	The name, now superseded, for the list of species from which the short and middle lists was derived.
Middle list	The name, now superseded, for the list of species in the second tranche of action plans.
Priority species or habitat	The current nomenclature for those species or habitats for which action plans have been prepared.
SAP	Species Action Plan.
Short list	The name, now superseded, for the first tranche of species action plans.
Species of conservation concern	The name now used for the database (formerly the long list) from which priority species are derived.
Species statement	A statement of action (typically search only) prepared for those priority species not seen within the last 10 years.
UK BG	UK Biodiversity Group.

Prepare action plans for threatened species in priority order: globally threatened; threatened endemics; other threatened species listed in the relevant schedules or annexes to UK and EC legislation and international agreements to which the UK is a party; endangered and vulnerable species listed in Red Data Books, aiming to complete and put into implementation plans for at least 90% of the presently known globally threatened and threatened endemics within the next ten years.

This action may have been stimulated to a degree by the publication of two editions of *Biodiversity Challenge: an Agenda for Conservation in the UK* (Wynne *et al.*, 1993, 1995), produced by a consortium of nongovernment organisations (NGOs). These documents contained a series of targets and summary action plans for a range of species considered to be priorities for action.

The Steering Group report

Subsequently, the Government established a Steering Committee to take forward the work of the Biodiversity Action Plan. This committee consisted of representatives of Government, of statutory conservation agencies and of NGOs. Their work culminated in the publication of *Biodiversity: the Steering Group Report* (Volumes 1 & 2) (Anon., 1995*a*,*b*). This report set the cornerstone for much of the work that has subsequently developed. In 1996, the recommendations of this report were broadly accepted by Government (Anon., 1996).

This response also initiated the creation of a network and hierarchy of groups to oversee the implementation of the action plans and the recommendations of the steering group report. The UK Biodiversity Group now takes responsibility for overseeing the development and implementation of the BAP. Four groups, one for each country in the UK, have been established to take the lead on matters relating to their geographic areas of responsibility. Additional groups lead on targets, on data (the Biodiversity Information Group), and on local issues and links to the Agenda 21 process (see Table 18.1), also arising out of Rio. Some of these groups have created their own subgroups so that the Scottish Biodiversity Group, for example, now has subsidiary groups dealing with costed action plans, public awareness, and local biodiversity action plans.

Species and habitat action plans

One of the most significant developments in the steering group report was the publication, in Volume 2, of action plans for 116 species (including four fungi), 14 action plans for key habitats, and statements for 37 broad habitats. The report also recommended that a further 290 species plans and 24 habitat plans should be produced within two years. As importantly, from a mycological point of view, it established the criteria for the further selection of species (and habitats) for action plan production. They are:

- that a species is globally threatened, or is possibly so; or

- that a species has declined by over 50% or more in the last 25 years.

Both the first 116 species (which became known as the 'short list') and the second tranche of species plans (known as the 'middle list') were selected from a list (the 'long list') of about 1200 species which were considered to be worthy of conservation attention. Subsequently, it was recognised that the long list, whilst lengthy, was by no means comprehensive and was an inadequate basis from which to select priorities for action. For instance, some species such as blue tits *Parus caeruleus* were included on the published long list simply because they appeared in an Appendix to an international convention, whilst many red list invertebrates and plants were excluded. Accordingly, the Joint Nature Conservation Committee (JNCC) was asked to organise a time-limited review of the middle list in late 1996.

However, this review was to be based on the agreed criteria without alteration, despite concerns expressed by some specialists on invertebrates and cryptogams that data were simply not available to enable these criteria to be satisfied. Although the revised middle list was subsequently larger than before, additional criteria were applied to limit the number of species subsequently to receive action plans: namely that any selected species should be taxonomically sound, that it had been recorded within the last 10 years, and that it would actually benefit from conservation action.

Since that review, it has been determined that all the species on the revised (and extended) middle list should be subject to some conservation action. Those species that met the subsidiary criteria above would still receive a full species action plan. The remainder would either be lumped, where possible, with another species or habitat action plan (if their requirements were the same) or, if they had not been recorded for over 10 years, the only action would be for targeted search. The latter, and any remaining species, would be the subject of a 'species statement'. The plans and statements were commissioned from a variety of sources including, for fungi and lichens, the Royal Botanic Garden Edinburgh and Plantlife, the UK wild plant conservation charity. Editing was overseen by a group of cryptogamic specialists from each of the Country Agencies and JNCC. These subsequent tranches of action plans and statements are mostly now published (Anon., 1999).

It has also been decided that the nomenclature of short, middle and long lists should be superseded. In future, any species that is the subject of an action plan will be referred to as a 'priority species'. The list from which these priorities have been determined will be referred to as 'species of conservation concern'. Indeed, this is now the title of a database, derived from the original long list, holding summary data on threatened species, which is maintained by the JNCC and managed by a consortium of country agency and NGO representatives. In due course, this database will be accessible via the JNCC web page (<http://www.jncc.gov.uk>), where copies of the species action plans can also be obtained from the UK Biodiversity website.

Implementing the action plans

In taking forward these action plans, a number of roles have been established. These include the following.

- *Contact point.* Always a Government agency, responsible for acting as initial contact for queries, stimulating action and maintaining standards, reporting progress, and confirming that the work programme meets the requirements of a published plan.
- *Lead partner.* Any organisation of conservation competence, responsible for co-ordinating the work programme with other partners, directing resources and stimulating publicity, establishing a partnership network to ensure that the work programme can be delivered, and reporting on progress.
- *Champions.* Any organisation or individual that wishes to support the work of an action plan by providing resources, in cash or in kind.

So far, no champions have been recruited to support the work of a fungus or lichen action plan. Plantlife have accepted the role of lead partner for many of the fungi and will be consulting with the mycological community. English Nature and Scottish Natural Heritage will lead on others. Habitat action plans are also assigned a *lead agency* that is always a part of Government or one of its agencies. The lead partners or agencies for these plans may establish a steering group to oversee and guide the implementation of a plan and to co-ordinate the, often high, number of contributors.

Local biodiversity action plans

The BAP steering group report also set in train the development of Local Biodiversity Action Plans (LBAPs). Typically led by Local Authorities, through the development of partnerships of relevant organisations and individuals, they aim to ensure that national priorities are taken into account at local level. As importantly, they seek to ensure that local

Table 18.2. *Fungi for which species action plans have been published*

Published 1995
Battarea phalloides
Boletus satanus
Poronia punctata
Tulostoma niveum
Published 1999
Armillaria ectypa
Boletus regius
Buglossoporus pulvinus
Hericium erinaceum
Hygrocybe calyptraeformis
Hygrocybe spadicea
Hypocreopsis rhododendri
Microglossum olivaceum
Stipitate hydnoid fungi[b]
Species statements (in preparation)
Boletopsis leucomelaena[a]

[a]Search only – not recorded since 1963; [b]a combined plan for all threatened species in the genera *Bankera, Hydnellum, Phellodon* and *Sarcodon*. Lists of all action plan species can be found at <http://www.jncc.gov.uk>.

priorities and concerns are also addressed. A key element of this process is that there is full involvement of local people. An audit of the biodiversity of the area under consideration is usually the first step in the process before going on to define priorities and to implement action. Following a number of pilot projects, there are now a significant number of LBAPs under way in the UK.

Discussion

How have fungi fared in all this?

Of the 116 original action plans, only 4 were for nonlichenised fungi (Table 18.2). A further 9 taxa have subsequently had action plans produced (with an additional single 'species statement'). Lichenised fungi have fared better with 7 species in the first tranche and a further 29 species in the second tranche. Indeed, of all the taxonomic groups, fungi have fared least well in terms of the proportion of the group receiving action plans (0.1% for fungi, 2.1% for lichens and 6.7% for birds) (Anon., 1998). Only 'arthropods other than insects' fare almost as badly. However, one may ask whether even this limited number of fungi deserve to be included based on a rigid application of the criteria. Often there is little evidence of decline in fungi due to a paucity of earlier records. Current or increased recording effort may paradoxically, even inevitably, provide evidence of an apparent increase in range or abundance (e.g. Lawson, 1999) even though the underlying trend may be one of genuine decline. For some of these species, especially the grassland *Hygrocybe* and *Microglossum* species, decline was inferred based on known losses of grassland habitat. Evidence of global threat or decline is generally still more poorly known unless a species is confined to Europe or it is a putative UK endemic. Indeed, the criteria as a whole may have placed an undue emphasis on the endemic biota of the UK rather than dealing with threatened species *per se*.

On the current criteria those fungi receiving action plans are, at best, going to be a token, and perhaps *ad hoc*, sample of priorities for conservation. It is likely that the selection of these fungi, from a broad range of possible candidates, will be the subject of debate for some time. Intuitively, mycologists may view the rarest fungi, or those with the most restricted distribution, as being most worthy of attention. Yet, typically, these species may already occur in protected areas, such as Sites of Special Scientific Interest. By contrast, action plans for species, such as *Hygrocybe calyptraeformis*, that are more widespread (but which were perhaps once even more common) will require the needs of fungi to be taken into account in management decisions for conservation over a much broader area. Consequently, they may have a greater impact on raising the profile of fungus conservation overall. Compare, for example, the distributions of the *H. calyptraeformis* and *Armillaria ectypa* (Anon., 1999), the latter being known from only a single site whilst the former is widespread though uncommon throughout the UK. Many of the vertebrates targeted in the BAP for action are also 'common' and widespread farmland, or even garden, species such as song thrush *Turdus philomelos* and skylark *Alauda arvensis*.

Regardless of the selection process, there is now a group of fungi, from a variety of habitats and of a variety of life strategies, which are firmly on the conservation agenda and which have started to receive targeted action by the conservation agencies and others (Table 18.3). For example, *Tulostoma niveum* has been subject to detailed studies of phenology and population at its, then, single known site in Scotland (Fleming, Ing & Scouller, 1998); a second site has since been discovered (E. Holden, pers. comm.). The Countryside Council for Wales (CCW) has commissioned studies of waxcap grasslands (Rotheroe, 1999; and see Chapter 10) and Scottish Natural

Table 18.3. *Examples of projects on fungi undertaken or commissioned by the statutory conservation agencies in the UK since the initiation of the BAP*

Agency	Project
Countryside Council for Wales	Waxcap grassland survey of Camarthen; identification of >50 sites for *Hygrocybe calyptraeformis.*
	Training on fungi, their identification and conservation.
	Production of leaflets and articles in media on fungus conservation.
	Database established on the rare fungi of Wales.
English Nature	Grant aid for survey of *Boletus regius* by Plantlife.
	Leaflet on collecting of fungi (1998).
	Contribution to waxcap survey.
Environment & Heritage Service (Northern Ireland)	National checklist of fungi of the British Isles (a partnership of all the Country Agencies and Royal Botanic Gardens Kew).
	Survey of Boletaceae and Amanitacaea of County Down.
	In-house surveys for waxcaps and *Microglossum olivaceum.*
Joint Nature Conservation Committee	Comparative survey of waxcap grassland fungi in Britain and Ireland.
	Fungi of National Vegetation Classification grassland communities.
	Collation and digitisation of foray records for the BMS database.
Scottish Natural Heritage	Survey of tooth (stipitate hydnoid) fungi in native pinewoods (1998–2000); including search for *Boletopsis leucomelaena.*
	Technical information on the occurrence of cryptogams in Scotland to assist the implementation of LBAPS.
	Phenology and monitoring of *Tulostoma niveum* at Inchnadamph.
	Waxcap grassland survey of Scotland 1999–2001.
	Inventory and database of fungi in the Cairngorms.

Heritage (SNH) is about to follow suit. SNH has also commissioned a study of tooth (stipitate hydnelloid) fungi in native pinewoods, English Nature have grant-aided Plantlife to undertake work on *Boletus regius* (Marren, 1998) and further work is planned. SNH and CCW have provided technical information on the occurrence of fungi by Local Authority boundaries and, in Scotland, by natural areas (Ward, 1999). It is no exaggeration to say that without the BAP process, the claim of fungi on the resources of the statutory conservation agencies, at least, would have been much smaller (even though the resources being committed are currently much less than those for other taxa).

What is the role of the British Mycological Society in implementing these plans?

Already, the British Mycological Society (BMS), through its waxcap grassland survey (Rotheroe *et al.*, 1996) is taking forward action that is fundamental to implementing these plans. Indeed, the BMS survey is recognised in the action plans for the two *Hygrocybe* species. All the plans will require good data if they are to be implemented effectively and the BMS database, and records collected by BMS members, will be vital in assisting those charged with taking plans forward. It is also recognised in this process that the selection of priorities is not set in stone. Better information in time may indicate that further species need to be considered as priorities for action. Alternatively, conservation or recording effort may indicate that a species no longer deserves the action being committed to it – a conclusion reached by Marren (1998) in his study of *Boletus regius*. Such reporting and feedback on progress with plan implementation are an integral part of the BAP.

Although this account has focused on the species action plans, there will also need to be an input by mycologists to the implementation of the Local and Habitat Action Plans if they are to achieve their potential to contribute to the conservation of fungi. Indeed, the maximum gains for the conservation of fungi, and of *all* fungi not just those selected for action plan production, may best be achieved by this approach. Many of the fungus action plans, and indeed those of other taxa, have been grouped by habitat preference to achieve synergy of effort (Simonson & Thomas, 1999). For example, implementation of the habitat action plan for wood pasture should benefit *Boletus regius*, *Buglossoporus pulvinus*, *Hericium erinaceum*, and some stipitate hydnoid fungi as well as other species such as the lichen *Enterographa sorediata* or the stag beetle *Lucanus cervus*. Other Habitat Action Plans cover ecosystems, such as sand dunes, native pinewoods, calcareous grasslands and wet woodlands, which are also rich in fungi yet the fungus interest will only be fully recognised with the active involvement of mycologists. Many Local Biodiversity Action Plans already include some fungi in their list of priorities.

Regardless of any faults in the process, the Biodiversity Action Plan has given a priority, momentum and resource commitment to the conservation of fungi in the UK that was inconceivable in its absence; it is now up to mycologists to make the most of the opportunities that have been presented to us.

Acknowledgements

I thank Maurice Rotheroe for comments on drafts of this article. I am also grateful to Stephen Ward (SNH), Nick Hodgetts (JNCC), Ray Woods (CCW), Dave Stone and Will Simonson (English Nature), and David Mitchel (Environment & Heritage Service) for the provision of information used in this chapter. The views expressed here are those of the author and do not necessarily reflect those of the Joint Nature Conservation Committee.

References

Anon. (1994). *Biodiversity: the UK Action Plan*. Cm2428. HMSO: London.

Anon. (1995*a*). *Biodiversity: the UK Steering Group Report.* Vol. 1, *Meeting the Rio Challenge*. HMSO: London.

Anon. (1995*b*). *Biodiversity: the UK Steering Group Report.* Vol. 2, *Action Plans.* HMSO: London.

Anon. (1996). *Government Response to the UK Steering Group Report.* Cm3260. HMSO: London.

Anon. (1998). *UK Biodiversity Group. Tranche 2 Action Plans. Volume I – Vertebrates and Vascular Plants*. English Nature: Peterborough.

Anon. (1999). *UK Biodiversity Group. Tranche 2 Action Plans. Volume III – Plants and Fungi*. English Nature: Peterborough.

Fleming, L. V., Ing, B. & Scouller, C. E. K. (1998). Current status and phenology of fruiting in Scotland of the endangered fungus *Tulostoma niveum*. *Mycologist* **12**, 126–131.

Lawson, P. (1999). Sandy stilt puffball. *British Wildlife* **10**, 361.

Marren, P. (1998). Boletus regius *Kromb., the royal bolete and the pretender* Boletus pseudoregius *Estades*. Plantlife Report No. 111. Plantlife: London.

Rotheroe, M. (1999). *Mycological survey of selected semi-natural grasslands in Carmarthenshire*. CCW Contract Science Report no. 340. Countryside Council for Wales: Bangor.

Rotheroe, M., Newton, A., Evans, S. & Feehan, J. (1996). Waxcap-grassland survey. *Mycologist* **10**, 23–25.

Simonson, W. & Thomas, R. (1999). *Biodiversity: Making the Links*. English Nature: Peterborough.

Ward, S. D. (1999). *Local Biodiversity Action Plans – Technical Information on Species: I. Cryptogamic Plants and Fungi*. Scottish Natural Heritage Review No. 70. Scottish Natural Heritage: Battleby, Perth.

Wynne, G., Avery, M., Campbell, L., Gubbay, S., Hawkswell, S., Juniper, T., King, M., Newbery, P., Smart, J., Steel, C., Stones, C., Stubbs, A., Taylor, J., Tydeman, C. & Wynde, R. (1993). *Biodiversity Challenge: an Agenda for Conservation in the UK*, 1st edn. Royal Society for the Protection of Birds: Sandy.

Wynne, G., Avery, M., Campbell, L., Gubbay, S., Hawkswell, S., Juniper, T., King, M., Newbery, P., Smart, J., Steel, C., Stones, C., Stubbs, A., Taylor, J., Tydeman, C. & Wynde, R. (1995). *Biodiversity Challenge: an Agenda for Conservation in the UK*, 2nd edn. Royal Society for the Protection of Birds: Sandy.

19

The Scottish Wild Mushroom Forum

ALISON DYKE

Introduction

The Scottish Wild Mushroom Forum was developed in response to the rapid growth of interest in harvesting wild mushrooms. Both picking for the pot and commercial harvesting are popular and bring people back into the countryside in an enjoyable and sometimes profitable way. The commercial industry, which has developed in the last fifteen years, involves five principal companies and up to 400 casual harvesters. Commercial harvesting is fairly important as it provides an additional source of income for rural communities.

Until recently there has been little mushroom harvesting in Scotland, and the traditions and widespread knowledge that exist on the Continent have not had time to develop. The Forum sees the enjoyment that people gain from harvesting mushrooms as an opportunity for positive action, and seeks to involve harvesters in the creation of a voluntary Code of Practice. As well as producing a Code, the Forum will seek to implement it in ways that will reach the largest audience. By involving people in the management of the resource that they use, the Forum intends to build a culture of sustainable harvesting.

Grants from Scottish Natural Heritage and Moray Badenoch and Strathspey Enterprise fund the Forum, together with a Millennium Award from the Millennium Forest for Scotland Trust.

The process

The groups involved in the harvest of wild fungi were identified during an earlier study (Dyke & Newton, 1999) and are shown in Table 19.1. Contact was made with these groups. In the case of mushroom harvesters announcements were made in the local press and at points of sale in three

Table 19.1. *Groups involved in harvesting*

Forum members	Landowners' representatives
	Mushroom Buyers
	Mushroom Pickers
	The Association of British Fungus Groups
	The British Mycological Society
	Cairngorm Partnership
	Forest Enterprise
	Grampian Fungus Group
	The Joint Committee for the Conservation of British Insects
	The National Trust for Scotland
	Scottish Native Woodlands
	Scottish Natural Heritage
Nonparticipating members	Plantlife
	The Royal Society for the Protection of Birds
	World Wide Fund for Nature
	Scottish Wildlife Trust
	Highland Birchwoods

Table 19.2. *Some of the issues surrounding wild mushroom harvesting*

Beneficial aspects of harvesting	Exercise and access to the countryside
	Money – the commercial aspect
Conservation aspects	Waste – picking and dumping of inedible or unwanted species
	Effects on colonisation
	Habitat disturbance – plants and animals
	Designated sites: Sites of Special Scientific Interest, Nature Reserves, etc.
Effects on estate business	Damage (deer fences cut, litter dropped)
	Stalking and cattle disturbance
	Permission
	Effects of land management on fungi
Competition between pickers	

areas where mushroom harvesting is concentrated (Deeside, Speyside and the Black Isle). Regional meetings were held where issues were discussed and representatives nominated. The issues the Forum aims to address are shown in Table 19.2.

Having established an agenda, Forum members were provided with background information to ensure that they were all aware of the concerns of other groups. A meeting of all representatives was held and issues that

Table 19.3. *The Scottish Wild Mushroom Code*

The countryside is a working landscape. Please be aware of safety and follow the Countryside and Access Codes. As a matter of courtesy you are advised to ask for permission before you pick mushrooms.

By respecting the natural environment you can help to manage and conserve the countryside. When picking mushrooms for any purpose, please consider the following points:

- Wildlife, especially insects, need mushrooms too, so **only pick what you will use**.
- Do not pick mushrooms until the cap has opened out and leave those that are past their best.
- The main part of the mushroom is below the surface, take care not to damage or trample it, and not to disturb its surroundings.
- Scatter trimmings discreetly in the same area as the mushrooms came from.
- Some mushrooms are poisonous and others rare and should not be picked – only pick what you know and take a field guide with you to identify mushrooms where you find them.
- Before you collect mushrooms at a nature reserve please always seek advice from the manager, as special conditions may apply.

If you own or manage land:

- Be aware that your management activities may affect mushrooms.

If you wish to run a foray or collect for scientific purposes:

- In order to ensure the safety of your party obtain permission in writing.
- Give a record of what you have found to the landowner or manager and explain the significance of your findings.

This Code was created by the Scottish Wild Mushroom Forum, a group consisting of representatives of: conservation organisations, landowners, public landowning bodies, mushroom buyers and mushroom pickers. The Forum was funded by Scottish Natural Heritage, The Millennium Forest for Scotland Trust and Moray, Badenoch and Strathspey Enterprise.

could be addressed by a Code were identified. In order to reach a variety of audiences the Code uses direct and positive language and will be produced in a number of formats.

The draft text of the Code is shown in Table 19.3 and will be supported by an informational leaflet. The first five points could be used as a set of 'golden rules' on signage, as a credit card sized handout, or even on wild mushroom packaging. It is hoped that educational materials for use by rangers and people wishing to run forays will also be available.

The Code will be endorsed by a number of organisations and will be distributed through a range of outlets in order to reach the widest possible audience. Plans are also being made for a pilot consent scheme, involving the issue of consents to mushroom harvesters. This would allow landowners to limit the number of commercial harvesters, to identify cars parked on their land and could also allow mushroom harvesters access to areas they would not normally be able to reach. A consent could also be used as a vehicle to introduce the Code to commercial harvesters.

It is intended that the process should remain open, with ongoing participation to review the Code and its implementation.

Reference

Dyke, A. J. & Newton, A. C. (1999). Commercial harvesting of wild mushrooms in Scottish forests: is it sustainable? *Scottish Forestry* **53**, 77–85.

20

The contribution of national mycological societies: establishing a British Mycological Society policy

DAVID MOORE

Introduction

The British Mycological Society (BMS) was founded in 1896 and today has about 2000 members who are located all over the world. The constitutional objective of the Society is to promote mycology in all its aspects by publications, meetings and such other means as it shall deem appropriate. The Council of the BMS is the executive that implements on a day-to-day basis the activity of the Society that is decided by the members of the Society at the Annual General Meeting, usually held in early December of each year.

Field mycology and an awareness and appreciation for the natural world were at the heart of the business of the Society from its foundation and these concerns continue as one of the Society's major activities today. The origins of the BMS trace back to the mid-nineteenth century (Webster, 1997). The Woolhope Field Naturalists' Club was based on the Hereford Museum, though members of the club dined at the Green Dragon Hotel in Hereford. In 1867 the Curator of the museum, Dr H. G. Bull, encouraged the club to take a special interest in fungi. He invited them to join him in 'a foray among the funguses' and this became an annual event, traditionally held in Hereford during the first week of October. The Woolhope Club meetings became a focus for all with an interest in fungi and attracted mycologists both from Britain and abroad. These forays lost their popularity when Dr Bull died in 1885 and stopped in 1892. By that time, though, the Yorkshire Naturalists' Union (YNU) was organising regular forays in different parts of Yorkshire and a Mycological Committee was formed in 1892. The stated aim of the YNU Mycological Committee was that their annual forays would take the place of the Hereford Foray and 'by avoiding the weak points of its predecessor, which were mainly confined to an excess of hospitality – prove at least equally attractive and instructive to mycologists' (Massee & Crossland, 1893, quoted by Ramsbottom, 1948). A need was also felt to provide an outlet for the publication of scientific articles on fungi. The idea of forming a 'National Mycological Union' emerged at the YNU meeting in Huddersfield in 1895 and the decision to set up the British Mycological Society was taken on 19 September 1896 at a meeting of the YNU Mycological Committee at the Londesborough Arms in Selby. The first officers were G. E. Massee as President (then Mycologist at the Royal Botanic Gardens, Kew), Charles Crossland as Treasurer (a Halifax butcher by trade, Crossland compiled, with Massee, *The Fungus-flora of Yorkshire* in 1902–1905) and Carleton Rea as Secretary. Rea was a barrister by profession but he gave this up in 1907 and wrote *British Basidiomycetae* (published 1922) which was the standard work on identification of the group for many years.

Autumn forays were the main activity of the Society initially and their arrangement was the chief responsibility of the Secretary. From 1919 a Foray Secretary took on this task. The first was A. A. Pearson (1919–24), who was a gifted amateur and an authority on the identification of agarics and boletes. He wrote popular keys to *Russula*, *Lactarius*, *Boletus*, *Inocybe* and *Mycena*, and in 1948, with R. W. G. Dennis, *A Revised List of British Agarics and Boleti*. The Annual General Meeting in 1942 established a Foray Committee. Later the committee was renamed the Foray and Conservation Committee.

Autumn, Spring and Day forays became a regular feature of the Society's annual programme. Regular specialist forays were inaugurated in 1982 with an Upland foray to Wester Ross. The first 'official' Truffle Hunt took place in the Cotswolds in 1984. They were enthusiastically continued for several years and included a truffle meeting in Italy in 1987. Regular annual hunts were discontinued because of the possible threat to rare fungi. Overseas Forays have been held in Northern France (1984), Northern Greece (1988), Southern Denmark (1991), Norway (1994) and are now also a regular feature of the programme. The first Tropical Expedition organised by the Society was an excursion to Cuyabeno, Ecuador, in 1992 (Hedger *et al.*, 1995). Collecting trips have more recently been made to Thailand and it is expected that tropical expeditions will be organised every three to four years. In 1998 the BMS held a joint meeting in Chiba, Japan, with the Mycological Society of Japan. Although not a field meeting, this sealed close contacts between the two Societies and will lead to further and closer intercontinental collaborations in the future.

Collections made during BMS forays and expeditions provided valuable information on numbers of species and species distributions: for example, of British truffles (Pegler, Spooner & Young, 1993). Lists of fungi collected on forays formed a regular part of Foray Reports in the *Transactions of the British Mycological Society* and the earlier volumes of the Society's *Bulletin*. With the establishment of a BMS Fungal Records Database in 1986, computerisation of records began (Minter 1986*a*,*b*) and that database is now becoming a major resource.

In the present committee structure of the Society there are several Special Interest Committees (SICs): Biodiversity; Ecology and Environmental Mycology; Foray; Fungus–Invertebrate Interactions; Conservation; Genetics, Molecular Biology and Evolution; Pathogenic and Mutualistic Interactions; Physiology; Systematics and Structure. The SICs for Conservation and for Biodiversity were set up in 1996 in recognition of the need for better focus on these particular aspects of mycology.

Developing the conservation agenda within BMS

Mounting concern over adverse effects of environmental and atmospheric pollution was reflected in data showing similar decline in fungal populations published from the 1970s onwards. This is not the place for a comprehensive set of references, but it is interesting to note the time span represented in the following few papers: Wilkins & Patrick, 1940; Wilkins & Harris, 1946; Richardson, 1970; Arnolds, 1988*a*,*b*; Eveling *et al.*, 1990.

In a report to Council dated 9 December 1986, the 1987 BMS President, Professor Roy Watling, reminded Council that 'In 1976 a list of rare larger fungi was discussed with BRC [Biological Records Centre] (Greenhalgh, Whalley and Watling representing BMS). It is possible to produce a list for Britain and a list of indicator species . . . '.

Subsequently, in April 1987, Roy Watling was even more instrumental in placing conservation firmly on BMS Council's agenda, in a perceptive document that stated: 'The British Mycological Society is committed to the conservation of our national heritage and to playing some defined and active role to meet these aims. The Society is in a good position to offer expert advice on individual sites and expert opinions on specific fungal records and this will improve in the future as more records are keyed into the Society's computer. To fulfil the above role, which is really one of communication and co-operation, the Society has already appointed a representative on the Conservation Committee of the European Mycological Congress, and its own Conservation Advisory Officer. It is hoped that as time goes on the names of these Officers will be more widely known, and their expertise called upon more frequently . . . Council should think in terms of setting up a small working party probably best linked to the Systematics, Structure and Foray SIC to consider how the proposed mapping of British Fungi can be used to monitor the decrease of particular species etc., and so find out the impact that foraying has on a selected fungus flora, or whether such a factor can even be measured.'

On 22 February 1988, Roy Watling made the specific suggestion that Council should finance some urgent research 'To circulate and collate information concerning fungal protection in European countries and to relate this to the situation in the British Isles.' The supporting documentation for this proposal clearly identifies the major concern being expressed at that time about 'the effects of collecting fungi, either for recreational or scientific purposes, on their productivity and, therefore, their long term continuance in nature. The Plant Protection Act has made a major contribution to the safeguarding of higher plants, but sadly the fungi were not included therein owing to lack of knowledge. Unfortunately, this lack of knowledge continues but there is an ever-increasing pressure on our fungi, including demands from abroad for edible species. Certain European countries already have a 'picking policy' and the British Mycological Society needs urgently to make a preliminary study of (a) the information which has brought this about, (b) the desirability of similar policies in the British Isles with all its attendant problems, and (c) the effects of introduction of such policies if considered necessary.'

Council agreed to support the project on 'Effects of picking fleshy fungi on the Countryside's resources' and the recently appointed Conservation Officer, Bruce Ing, was able to report to Council in July, 1988, details of the programme of research and that the work would be undertaken by Dr Thomas Læssøe, who then reported in November of the same year.

As far as I am aware the Læssøe report *Conservation of Fungi* remained a Council discussion paper which was never published, but as its findings formed much of the foundation for subsequent BMS activity it is worth giving some extensive quotations from it. The first few paragraphs established the history and background of the study.

> Since the invitation to European mycologists by the Dutch group to form a committee on fungal conservation [at] the European Mycological Congress in 1985 the subject has become topical. Many journals including the *Mycologist* and *The Transactions of the British Mycological Society* have published papers on the protection of fungi and have pointed to alarming decline in some fungal groups in specific areas. Acid rain, over-picking, bad management, etc., have been put forward as reasons for the alleged

decline. The Council of the British Mycological Society also felt that something should be done and appointed an officer to the European Committee (Roy Watling) and an officer to deal with the local matters (Bruce Ing). These two officers together with the General Secretary (Tony Whalley) became the board for my project: To write a summary of the situation and give some suggestions of how to proceed.

History of mycological conservation

For obvious reasons the direct conservation of fungus species have been a very insignificant part of general conservation schemes and only recently have fungi been taken into consideration by conservationists.

In the USSR all fungi included on the Red Data List are automatically protected. In Poland a special list of protected fungi have been prepared and e.g. advertised to the public by issuing a stamp series with pictures of the different species. In some countries (e.g. the German Federal Republic) it is prohibited to market certain edible but rare species.

A more important thing for a long-term conservation purpose is of course to protect sites rather than species. But again very few sites have been protected because of a known important mycoflora. There is only one nature reserve in the UK created to protect the mycoflora (a 'hedgerow-locality' with *Battarrea phalloides*). One or two other sites have been listed as sites of special scientific interest because of an interesting and well-documented mycoflora. Protection of sites because of mycological interests is equally rare in other parts of the world. . . .

Possible causes of fungal decline (or increase): Picking . . . Although it is commonly believed that mushrooms are threatened by overpicking, no scientific data support this view (Jansen & van Dobben, 1987, Arnolds 1988[b], etc.). Fruiting of some species seems to be enhanced by cropping and it has been postulated that hypogeous fungi are favoured by soil disturbance. Nitare (1988) warns against continued picking of fungi already recorded in an area although the taxonomic difficulties often make it necessary to collect specimens. In Oregon a project [has] recently been set up to study the cropping of *Cantharellus cibarius* over a 10 year period. The aim is also to measure the effect of trampling (mechanical damage to the soil and supporting vegetation). This might be a more severe threat than the actual picking. Arnolds (1988*b*) made a seven year study where fungal fruit bodies were removed from plots. No decrease in fruiting was observed.

. . . The British Mycological Society policy concerning picking is somewhat unclear, since on [the] one hand R. Watling in *CAB News* and in the *Mycologist* advises the membership to collect only what is necessary for a determination while in each issue of the *Mycologist* you will find instructions of how to cook (etc.) your mushrooms. In most European countries, with the exception of The Netherlands, it would be considered completely non-desirable to condemn picking for culinary purposes.

Land management

"The principal reasons for the decline of fungi are forest management practices, including the reduction in the numbers of rotting trees in commercial forests, the artificial drainage of wetlands, the encroachment of forest and especially spruce trees on what was originally meadowland, and constructions." (from the Finnish Red Data Book).

This statement is almost universal in the different conservation papers. Another general factor is the agreement on the negative effect of commercial fertilisers both in grasslands and, in forests. Even coprophilous species such as some *Coprinus* species and *Poronia punctata* are negatively affected by adding commercial fertilisers (Arnolds, 1988*b*).

The removal of dead trunks and larger fallen branches is also considered a general problem.

In a table [unpublished] communicated by Bruce Ing the loss of major habitats in the UK since 1949 is listed. Only 3% of 'lowland neutral grasslands' remain undamaged and only 15% of 'ancient lowland woods of native broad leafed trees' remain undamaged . . .

Pollution

Both Arnolds (1988*b*) and Nitare (1988) (and various Dutch and German papers) stress the importance of airborne pollutants in fungal decline. In all probability airborne pollutants are responsible for the decline of ectomycorrhizal fungi in The Netherlands, exemplified by detailed studies on the Chanterelle. Jansen & van Dobben (1987) stated about the decline of *Cantharellus cibarius* "although our data indicate an effect of acidification or eutrophication, effects caused by heavy metals are also possible. Heavy-metal pollution may also lead to an increase in the accumulation of organic matter. In one case there is evidence of sensitivity of *C. cibarius* to heavy metals (a Swedish study)". In the Dutch material levels of heavy metals were lower than in the proven cases of toxicity, and thus heavy metals were less likely to be the cause of the decline.

Judged by the communications in Poland also the situation in Czechoslovakia and in parts of Germany is very grave concerning the ectomycorrhizal fungi, while the situation in Scandinavia is either not documented or much less severe. The same seems to apply for the UK.

Climatic changes

Many fungi considered rare in various countries can probably best be regarded as being outposts from their natural distribution and they are thus sensible to even small changes in the climate. No doubt some of the disappearing species can be regarded as 'threatened' by climatic changes (e.g. several elements in the south eastern British mycoflora). The annual variation in climate no doubt also is responsible for many records of so called rare fungi, e.g. in this summer in Denmark where the unusually hot and humid July has resulted in a number of records of "rare, southern" species.

In reaching his conclusions, Læssøe referred to a paper by Kirby (1988) from the Chief Scientific Directorate of the Nature Conservancy, who raised three questions about fungal conservation: (A) are special conservation measures for fungi needed or justified? (B) How should we judge which sites should be conserved for their fungi? (C) How should these sites be managed to maintain their fungal value?

Læssøe concluded that special conservation measures for fungi are justified, that Red Data lists can provide the basis for deciding which sites are most important, and recommended that site management should ensure, for example, that nutrient poor grasslands should be kept nutrient poor (by avoiding fertilisers) and should be grazed or cut; old wood should be left in forest reserves, and drying of the topsoil should be avoided.

A number of specific projects for BMS to undertake were proposed in the Læssøe report. These were, in the order of preference quoted by Læssøe: production of a booklet on conservation of fungi; production of a code of conduct leaflet for forays and mushroom pickers; updating of the BMS database so it can help in preparing Red Data lists; protection of valuable grasslands (including dunes); study of airborne pollution and its effect on the mycota; and issuing questionnaires to the membership to assess change in distribution of indicator species.

Also in November 1988, Roy Watling reported to Council a successful meeting on Fungal Conservation held on 12 November which was organised for the BMS by David Minter. Delegates at the meeting included Dr A.-E. Jansen, the Secretary of the European Conservation Committee, Dr N. Stewart from the Conservation Association of Botanical Societies, and Dr N. Hodgetts and Dr Keith Kirby, of the Chief Scientific Directorate, of the Nature Conservancy Council. This was the first in a number of scientific meetings on this topic area sponsored by the Society. Later ones included a Symposium on Fungi and Environmental Change held in 1994 (Frankland, Magan & Gadd, 1996), and the Symposium at the Royal Botanic Gardens, Kew (13 November 1999) entitled Fungal Conservation in the 21st Century, which was the origin of this book.

All the above shows the careful scientific approach expected of a learned society. There is a need for cogent logical arguments if we hope to change what people are doing, but above all there is a need for relevant scientific knowledge.

In the course of the few years immediately following the Læssøe report the Society's then Conservation Officer, Bruce Ing, published a provisional Red Data list of British fungi (Ing, 1992); he has revised and re-evaluated these data on a regular basis since then. However, the first formal outcome of discussions prompted by the Læssøe report was published as the ten point BMS policy on conservation (Anon., 1990), stating:

1. The British Mycological Society is committed to the conservation of non-harmful fungi and their habitats. To this end it will foster and support those activities which will ensure the survival of fungal populations.
2. The Society will compile a Red List, using strict criteria and related to habitat type, of rare and endangered fungal taxa, which may be helpful in evaluating sites and in drafting any future legislation on species protection.
3. The Society will, where possible, seek to provide information concerning the mycoflora of sites which may be threatened, so that management appropriate to fungal conservation may be planned. The Society will provide advice to landowners, conservation bodies and local authorities on the mycological importance of their land and of suitable conservation measures that may be adopted.
4. The Society will encourage research into the decline, or otherwise, of fungal populations and will help to make available the results of such research. The Society's database of fungal observations will assist in this motoring activity.
5. The Society maintains that there is no evidence that the responsible collection of fungi for scientific purposes presents any threat to populations and, moreover, that it is essential for the accurate identification of species and the compilation of site inventories.
6. The Society will provide a Code of Conduct for its members, other conservation bodies and individuals who wish to collect fungi.
7. The Society does not condone commercial collection of wild mushrooms, particularly as there is uncertainty as to the effects of such large-scale collection on fungal populations.
8. The Society will strengthen its links with those organisations concerned with forest ecosystems, especially where such systems are seen to be under stress, so that the vital role that fungi play in such ecosystems becomes more widely appreciated.
9. The Society will extend its relationship with its European and other overseas counterparts to exchange information concerning continental and global changes in mycoflora, and will encourage

appropriate research programmes. It will review and update the policy regularly so as to take into account new research findings and policy decisions in other countries.

10. The Society will actively promote a wider understanding of the importance of fungi and their biology, and in particular their significance in the conservation of natural communities, as a contribution to environmental education.

During the next few years the Society concentrated on codifying its conservation policies, with particular emphasis on collaboration with other bodies, both national and international, in an effort to increase and widen the effectiveness of those policies.

Codifying policy

At a meeting of Foray Group Leaders convened by the Society at Littledean in May 1995 Maurice Rotheroe, then speaking as Deputy Conservation Officer (and Conservation Officer-elect) expressed the view that the Society could make a much greater contribution to local and national conservation issues and there was a need for conservation to have a much higher profile within the BMS. He outlined the role that the BMS could play in liaison with English Nature, the County Trusts and other wildlife and conservation bodies. The BMS could offer specialist knowledge of the relationship between fungi and other organisms not available elsewhere. There was much work to do in publicising fungi, in educating reserve managers and others in mycology and management for conservation of fungi. He also issued a questionnaire to Group Leaders to enable him to collect the first-hand experience of these groups and to establish links with local County Trusts and English Nature representatives in each area.

Minutes of the BMS Council meeting of 6 December 1996 featured the annual report to Council of the BMS Conservation Officer, Maurice Rotheroe. This was especially notable because in the course of 1996 a Conservation Special Interest Committee had been established. The SIC's first meeting lasted four hours during which its members took on responsibility for gathering more data on endangered species to contribute to future Red Data lists; initiated work on the production of a draft guide to site management techniques which are sensitive to the requirements of fungi; and identified commercial collecting of wild fungi as the most prominent current issue following intense publicity about the topic earlier in the year.

On behalf of the Conservation SIC, Patrick Leonard compiled a consultation document entitled *Outline Policy on Commercial Collecting of Wild Fungi* which was presented to Council at this December 1996 meeting. This report notes that the BMS had set out its policy on commercial collecting of macrofungi as part of its policy on conservation, published in 1990 (Anon., 1990; and see above). Public concern had mounted as commercial collecting had expanded in the early 1990s and the policy needed amendment. The vehemence of the concern among some members of the public had been brought home to the Society in a face-to-face confrontation several weeks before this Council meeting. 1996 was the BMS Centenary year and at the Society's 'Fungus 100' exhibition in London in 1996, a wild-mushroom cooking demonstration by one of the world's great chefs, Antonio Carluccio, was noisily interrupted by 'green activists' accusing him of pillaging the environment!

Increased collecting of fungi gave rise to concerns about: (1) the effect on the long-term survival of the fungus, and the indirect effect on organisms further down the food chain, that is, the overall wildlife conservation issue; (2) potential distortion of scientific records; (3) the effect on local people who thought they had *de facto* collecting rights in public forests; (4) effects on landowners, and particularly woodland owners, of the nuisance of ever more determined pickers; (5) adverse aesthetic effects on the natural environment caused by indiscriminate collecting in autumn woods.

Patrick Leonard argued that the then published position of the BMS appeared weak and difficult to explain to concerned but nonexpert members of the public. It also failed to address the question of the cause of observed declines in fungal populations because point 7 in the 1990 policy statement (see above) simply refers to 'uncertainty as to the effects of such large-scale collection'. Inevitably, also, the 1990 policy had no reference to the International Convention on Biological Diversity that the UK signed in 1992.

Leonard suggested that 'A much more viable policy position would be to acknowledge that there are declines in the populations of common edible fungi. To point clearly to habitat losses and the effects of pollution and make clear the BMS stance on these. To adopt the precautionary principle in relation to commercially collected species where numbers are declining rapidly and propose to Government that action plans should be drawn up for two or three of these. To protect habitats where rare and threatened fungi are present by halting commercial collecting. To make common cause with the conservation agencies, the Forestry Authority, landowners and with the major conservation voluntary bodies such as the

National Trust and the Wildlife Trusts . . . In the light of increasing concern about the rapid expansion of commercial collecting of fungi in Britain it would seem desirable for the BMS to re-examine and strengthen its conservation policy. Such a change would be seen as a positive response to the concerns and a timely updating of the policy following new national and international policies on conservation . . . '.

Council agreed on the need for a revised conservation policy and that there should be a statement which presented the Society's views on the commercial collecting of wild fungi. The President (John Webster) asked Maurice Rotheroe to receive and collate contributions to a revised policy document which would be brought back to Council before being published in the *Mycologist*.

Consultations continued through 1997 and the final policy document was approved at the December 1997 meeting of Council, with David Moore as President. The text, published two months later (Anon., 1998a), was as follows.

British Mycological Society Policy on Conservation

1. General Statement:The British Mycological Society is committed to the conservation of fungal populations and communities. To this end it will foster and promote those activities that contribute to survival of viable fungal populations and communities.
2. Threatened Fungi: The Society will compile and publish a Red List that conforms to international standards and will press for positive conservation of threatened fungi through national and international measures.
3. Habitat Conservation: The Society will seek protection for important mycological sites against loss, deterioration or fragmentation, howsoever caused.
4. Edible Fungi: The Society acknowledges the importance of edible wild fungi as a resource to be utilised, but accepts harvesting of such fungi only where it is non-threatening to the viability of fungal populations, and their associated organisms and habitats.
5. Code of Conduct: The Society will publish a Code of Conduct for the responsible collecting of fungi.
6. Research: The Society's constitutional object 'to promote mycology' encompasses encouragement of research on the biology, including taxonomy and ecology, of fungi; on the causes of decline of fungal populations; and on the cultivation of edible fungi.
7. Information: The Society will monitor and record the occurrence of fungi and make its Database available to its members and to outside organisations and individuals.
8. Education: The Society will promote a wider understanding of the importance of the conservation of fungi.

9. Collaboration: The Society will enhance its links with organisations concerned with conservation and the protection of the environment at local, national and international levels.
10. Review: The Society will review and update its conservation policy, as required, to take account of new research findings and changes in relevant legislation and environmental policy.

The published policy was accompanied by a paragraph describing how practical implementation of the BMS Conservation Policy would be incorporated into a Five-year Plan which the Conservation SIC had developed and had been approved in principle at Council. As recorded in Council Minutes, the consultation document described the provisional five-year plan in the following terms.

1. *General:* The British Mycological Society will disseminate its policy on the conservation of fungi both inside and outside the scientific community, using appropriate media including publication via the World Wide Web.
2. *Threatened Fungi:* The Society will review and update the Provisional Red Data List of British Fungi and will press for national and European legislation to protect threatened fungi where appropriate. It will urge the Joint Nature Conservation Committee to publish a Red Data Book for fungi and will assist in its production. The Society will actively participate in the implementation of action plans for fungi published under the UK Biodiversity Action Plan.
3. *Habitat Conservation:* The Society will co-operate with statutory and voluntary conservation organisations to identify mycologically important sites and habitats. It will compile a management handbook, covering all ecosystems, to assist landowners and managers.
4. *Edible Fungi:* The Society will monitor the impact of commercial collecting and will press for regulation where such collection is shown to threaten the viability of fungal populations and their associated organisms. It will actively discourage the collecting of species that are identified as being under threat. The Society will co-operate with land managers and other bodies to promote the sustainable harvesting of edible fungi. It will initiate action through the European Council for the Conservation of Fungi to urge governments of importing and exporting countries to curb excesses of commercial collecting through legislation.
5. *Code of Conduct*: The Society will seek sponsorship for the production, publication and promotion of its Code of Conduct.
6. *Research:* The Society will press for research relevant to conservation of fungi and will compile a list of key areas for investigation, to be distributed to grant-awarding and research organisations in the public and private sectors. Specific areas for priority investigation will include: the impact of commercial collecting on fungal populations; the cultivation of mycorrhizal species of edible fungi; and the effect of

government agricultural and land-use policies on fungal populations and communities.

7. *Information:* The Society will ensure that its database is efficiently managed and updated and the design modified, where necessary, to make it compatible with similar databases held by other conservation agencies. It will place a high priority on participation in the National Biodiversity Network. The Society will encourage the exchange of data on fungi with European mycological societies.
8. *Education:* Officers of the Society will publicise the conservation activities of the Society, will promote educational training in fungal conservation and encourage other bodies to include consideration of fungi in their conservation and management activities. The Society will press for a programme of training to be set up for commercial collectors, restaurateurs and other traders in wild fungi in order to encourage responsible collecting and consideration of the importance of the long-term sustainability of the resource.
9. *Collaboration:* The Society will work with the European Council for the Conservation of Fungi, the International Union for Conservation of Nature and Natural Resources and the Berne Convention to seek protection for mycological habitats of European importance through the implementation of appropriate government land-use policies. Opportunities will be investigated for the financing of international fungus conservation research programmes. The Society will organise an international symposium in 1999 in order to review the future of fungal conservation in the 21st century.
10. *Review:* The Society will encourage its members to inform the Conservation Special Interest Committee of publication of new research relating to fungal conservation, of impending changes in legislation on environmental matters and of any modifications of policy by statutory or voluntary conservation agencies.

Consultations continued through 1997 and 1998 with English Nature, The Woodland Trust, The National Trust and the Forestry Commission, leading towards the publication of two leaflets entitled *The Wild Mushroom Pickers Code of Conduct* and *The Conservation of Wild Mushrooms* (Anon., 1998*b*,*c*).

The BMS input to these leaflets, which was crystallised over many years from the contributions of numerous individuals and committees can be paraphrased as follows.

British Mycological Society: collecting fungi from the wild, a code of conduct

Beauty, intrigue and value. All these can be found amongst the fungi. Exotic displays of mushrooms and toadstools in our woodlands and pastures, particularly during the autumn, are as much a part of our natural heritage as the more commonly appreciated displays of wild flowers. And the flushes of fungi can provide as much aesthetic appeal as their cousins, the flowers.

But fungi are also important for their medical and industrial value. They are sources of life-saving drugs and other products in daily use by us all. But more than this: fungi are essential components of natural habitats like woodlands, meadows and pastures and without them the ecology of these places could not function.

Fungi of all sizes are used as food by a range of animals. Mammals like squirrels, voles, deer, and we, of course, take the larger fruit bodies. Indeed, these provide homes as well as nutrient to many invertebrates (worms, insects, slugs), some of which may be rare or endangered species in their own right.

We collect fungi too. Some for scientific study and identification; some for screening for new drugs; but most, perhaps, for food use by the collectors themselves or for sale to groceries and restaurants.

We treat wild fungi as a natural harvest, the hunter-gatherer's perquisite! Which is all well and good, but humans don't have a particularly good track record for reasonable exploitation of such natural resources. The intention of this Code is to minimise the adverse impacts of our collecting activities. We want to maximise enjoyment of fungi whilst minimising the damage we might cause to other wildlife and wild places, to fungal populations in nature and, indeed, to landowners, site managers and other collectors.

Need for a code

Mycologists (people who study fungi) have known for many years that populations of wild fungi are subjected to the same adverse pressures as are other wildlife by our modern life style. This has led to the compilation of Red Data Lists of threatened species in the hope that recognising the threat will assist in their protection.

Right across Europe there has been a decline in the fruiting of some wild fungi. A particularly worrying aspect of this is that it is especially true for mycorrhizal mushrooms – which form an intimate symbiotic association with living tree roots – and which cannot yet be cultivated. Two popular edible mushrooms belong to this category: the Cep, *Boletus edulis*, and the Chanterelle, *Cantharellus cibarius*. The main causes of the decline appear to be loss of habitat and air pollution, and edible and inedible species of mushroom are equally affected.

But there are other pressures. Increasing popularity of 'field mushrooms' and 'wild fungi' on the menus of fashionable restaurants and in recipes of fashionable cookery books has increased the profitability of collecting from native populations. Collecting for commercial gain has become more common and is often concentrated on a few areas known to yield good harvests and has led to inevitable resentment and conflict with local residents and landowners.

The aim of this Code is to recommend good practice for all those who collecting fungi from the wild.

Principles

A number of common-sense principles underlie this Code:

- Scientific research has not detected any ecological damage arising directly from harvesting fungi, even on a large scale for commercial purposes. BUT, such studies are in their early stages. What IS clear is that the act of collecting can cause collateral damage by the effects of trampling, disturbance, removal and even destruction of natural areas.
- Mushrooms, toadstools and other fruiting structures are only the visible spore-producing bodies of the fungus, like the fruits of our orchard trees. The bulk of the fungus exists below ground as the fungal mycelium. Removing a mushroom may do no more damage than taking an apple from the tree (even though dispersal of spores, like discharge of plant seeds, is necessary for sexual reproduction). But trampling the mycelium to death can cause untold damage to the existing population.
- While we are so ignorant of the biological requirements of many fungi are not, it would be prudent to be cautious in our exploitation of natural resources. We must not risk long-term damage by ignorant exercise of our 'right to collect'.
- Nevertheless, wild fungi must be seen as a legitimate resource to be harvested. But we must ensure that harvesting is sustainable and does no damage either to the fungal population or to populations of organisms associated with the fungi.

Scientific study of fungi, such as recording their occurrence, sometimes requires removal of fruit bodies, even when the study is aimed at conservation. The reason is that proper identification may need microscope observations, or even molecular analyses.

Fungi and the law

Two main laws may protect fungi.

The Theft Act (1968), which applies only to England and Wales, makes it an offence to:

- Dig up and take any plant, tree or shrub or any soil etc. which is part of the land, being the property of somebody, so digging up fungi could constitute theft unless the landowner has given permission.
- Take the property of somebody and sell it for gain. The emphasis here is on selling for gain. The custom of taking wild fruit and flowers, including fungi, is permitted by the Act so long as there is no personal financial gain. Sale of collected fungi without the landowner's permission may be an offence.

The Wildlife and Countryside Act (1981), and a similar law in Northern Ireland, makes it an offence to:

- Pick, uproot, destroy or sell, and/or collect and cut any plant listed on Schedule 8 attached to the Act – and that schedule now includes some species of fungi.
- Uproot any wild plant, unless the person is authorised.

Remember also that special protection restricting collection of fungi may apply to Sites of Special Scientific Interest (SSSIs), designated under the Wildlife & Countryside Act, and National Nature Reserves (NNRs).

Local Bylaws forbidding the picking of fungi and plants on National Trust property, Crown Estates, Local Nature Reserves, Forestry Commission and Local Authority land may also apply.

The fungus pickers' Code

General guidelines

Many of the recommendations below are simply common courtesy or countryside etiquette. Remember that we share the environment with many other organisms and interests and so are the custodians of our natural heritage.

Before entering any land, get the landowner's or site manager's permission, explaining the purpose of your visit. Check for Bylaws concerning picking of fungi. Collecting is not allowed on some SSSIs, Nature reserves and other protected areas.

Follow the Country Code

Be sensitive to the structure of the natural habitat. Do not damage vegetation or soil, nor disturb unnecessarily leaf litter or other features. Do not move or remove dead wood unless essential to identify a fungus.

Collecting for food

Be VERY aware that some fungi contain deadly toxins and many more may make you unwell. Some people suffer allergic reactions after eating particular fungi, even though they may be well known as 'edible'.

Many mushrooms can concentrate heavy metals so do not collect in heavily polluted areas. This warning applies to the verges of busy roads (lead pollution) and to reclaimed land sited on old landfill waste dumps (where the danger is of industrial heavy metals like cadmium and mercury).

Do not collect large numbers of specimens that you don't recognise on the off chance that some might be edible.

Some edible species have poisonous look-alikes. Beginners should not eat anything that has not been checked by an experienced field mycologist. And remember that mainland Europe, where amateur mushroom hunting is a popular pastime, has the highest reported incidence of mushroom poisoning in the world!

Respect, and protect, all species, including poisonous ones.

Never collect rare or Red Data list species – and certainly not the species that are scheduled in the Wildlife and Countryside Act (listed below).

Even if the population is plentiful, take no more than you need for your personal consumption or use.

Collecting with a view to selling for profit or other commercial use must be agreed beforehand with the landowner.

Generally, never take more than half of the fruit bodies you can see.

Generally, never collect unopened or 'button' mushrooms. Giving the fruit bodies time to expand will allow spores to be discharged and result in bigger, perhaps tastier, mushrooms to be picked later.

Scientific collecting

Collect the minimum amount of material or number of specimens required for proper description and reliable identification.

Minimise the disruptive effects of taking samples. For example, replace rolled-over logs and avoid the unsightly damage caused by cutting chunks out of long-lived bracket fungi.

Always offer the results of your fieldwork to the landowner or site manager, with explanation of the significance of your findings.

Record localities and habitat data for rare species accurately and retain dried 'voucher specimens' for deposit in herbarium collections. Supply information to local and national databases. Remember that science needs to be communicated!

If you have permission to collect for scientific purposes do not abuse it by collecting edible fungi for eating 'on the side'!

The following fungi are protected under the Amendment to Schedule 8 of the Wildlife & Countryside Act, 1981:

Sandy Stilt Puffball, *Battarrea phalloides* – inedible.
Royal Bolete, *Boletus regius* – edible.
Oak Polypore, *Buglossoporus pulvinus* – inedible.
Hedgehog Fungus or Lion's Mane, *Hericium erinaceum* – edible.

. . . but you can't please all of the people all of the time

One of the most remarkable, and unexpected, aspects of the reaction to publication of the *Code of Conduct* was a thunderous editorial in the *Daily Telegraph* of 31 August 1998 headlined 'Mushrooming bureaucracy'. An accompanying article by the newspaper's Environment Editor, Charles Clover, gave a fairly measured account, though it bore the challenging headline 'Country code to curb the mushroom picker'. However, the editorial started off 'After Greenpeace, green police? This week will see the publication of one of those increasingly familiar 'codes of conduct' which tell you what to do in areas where we have managed well enough without rules. This one is to control mushroom picking, for heaven's sake, and it promises to provide splendid new opportunities for minor-league job-sworths whose hobby is bossing us about.' This gem of the journalist's trade ended 'Leave us to pick our mushrooms in peace, and if we wish to sell them, that is our own affair.'

In some quarters, then, the *Code of Conduct* might well be seen as overly challenging, but this could reflect the attitudes of at least a few of those involved in bringing together the contributions of different organisations. Certainly, during the discussion phase a representative of one of the partner organisations with which we were trying to establish a joint compromise approach e-mailed the BMS Conservation Officer stating 'I would remind you that much of your text . . . is the intellectual property of K--------, M----- and I, and you do not have our consent to use it . . . ' – not a very helpful stance for one's co-authors! Nevertheless, careful political navigation around the jealously self-protective (the 'jobsworths' in the eyes of the *Daily Telegraph*'s Editor) can, and in this case did, lead to a satisfactory conclusion – a document agreed by all parties concerned which could actually help the situation.

Dissemination of the advice and continued education is the key to further progress. It is clearly not good enough to limit our targets to the mycological community. The message must be spread to other concerned naturalists and then to the wider lay audience, especially managing and legislating authorities. This is something a national Society can, and should, take into its responsibility.

References

Anon. (1990). British Mycological Society policy on conservation. *Mycologist* **4**, 52.

Anon. (1998*a*). British Mycological Society policy on conservation. *Mycologist* **12**, 35.

Anon. (1998*b*). *The Wild Mushroom Pickers Code of Conduct*. English Nature: Peterborough, UK.

Anon. (1998*c*). *The Conservation of Wild Mushrooms*. English Nature: Peterborough, UK.

Arnolds, E. (1988*a*). Dynamics of macrofungi in two moist heathlands in Drenthe, The Netherlands. *Acta Botanica Neerlandica* **37**, 291–305.

Arnolds, E. (1988*b*). The changing macromycete flora in The Netherlands. *Transactions of the British Mycological Society* **90**, 391–406.

Eveling, D. W., Wilson, R. N., Gillespie, E. S. & Bataillé, A. (1990). Environmental effects on sporocarp counts over fourteen years in a forest area. *Mycological Research* **94**, 998–1002.

Frankland, J. C., Magan, N. & Gadd, G. M. (1996). *Fungi and Environmental Change*. Cambridge University Press: Cambridge, UK.

Hedger, J., Lodge, D. J., Dickson, G., Gitay, H., Læssøe, T & Watling, R. (1995). The BMS expedition to Cuyabeno, Ecuador: and introduction. *Mycologist* **9**, 146–148.

Ing, B. (1992). A provisional Red Data list of British Fungi. *Mycologist* **6**, 124–128.

Jansen, E. & van Dobben, H. F. (1987). Is decline of *Cantharellus cibarius* in The

Netherlands due to air pollution? *Ambio* **16**, 211–213.
Kirby, K. J. (1988). The conservation of fungi in Britain. *Mycologist* **2**, 5–7.
Minter, D. W. (1986*a*). Fungus recording. Computerization of foray records. *Bulletin of the British Mycological Society* **20**, 34–38.
Minter, D. W. (1986*b*). Fungus recording. Foray records database. Further information for users. *Bulletin of the British Mycological Society* **20**, 101–105.
Nitare, J. (1988). Skydd av hotade svampar, svensk och internationellt arbete [Protection of endangered fungi, Swedish and international work]. *Jordstjärnan* **9**, 25–33.
Pegler, D. N., Spooner, B. M. & Young, T. W. K. (1993). *A Revision of British Hypogeous Fungi*. Royal Botanic Gardens: Kew.
Ramsbottom, J. (1948). The British Mycological Society. *Transactions of the British Mycological Society* **30**, 1–12.
Richardson, M. J. (1970). Studies of *Russula emetica* and other agarics in a Scots pine plantation. *Transactions of the British Mycological Society* **55**, 217–229.
Rotheroe, M. (1998). Wild fungi and the controversy over collecting for the pot. *British Wildlife* **9** (6), 349–356.
Webster, J. (1997). The British Mycological Society, 1896–1996. *Mycological Research* **101**, 1153–1178.
Wilkins, W. H. & Harris, G. C. M. (1946). The ecology of the larger fungi. V. An investigation into the influence of rainfall and temperature on the seasonal production of fungi in a beechwood and a pinewood. *Annals of Applied Biology* **33**, 179–188.
Wilkins, W. H. & Patrick, S. H. M. (1940). The ecology of the larger fungi. IV. The seasonal frequency of grassland fungi with special reference to the influence of environmental factors. *Annals of Applied Biology* **27**, 17–34.

21

The contribution of national mycological societies: The Dutch Mycological Society and its Committee for Fungi and Nature Conservation

MARIJKE M. NAUTA & LEO M. JALINK

The Dutch Mycological Society

The Dutch Mycological Society was set up in 1908 and the object of the Society is to promote the study and understanding of fungi. This is done by organising meetings and field excursions, and by publication of the Society's journal, *Coolia*. The Society has a number of Committees, of which the Committee for Fungi and Nature Conservation deals with all aspects of conservation of fungi and nature management.

Apart from the yearly meeting in which all matters concerning the organisation of the Society are discussed, two general days with lectures are organised each year: one with a special theme organised by the Scientific Committee and a more general one organised by the Society itself. These meetings are very well attended, and at the general meeting in January it is often the case that 170 of the approximately 500 members attend.

In the mushroom season at least every weekend and often also during the week field excursions are organised for members. Every year two week-long forays are organised, one inland and one abroad. A good policy concerning conservation of fungi is impossible without knowledge of their distribution. Therefore, we will first briefly discuss the activities of the Dutch Mycological Society in the study of fungal distribution.

Distribution of fungi: recording

The Society is very active in gathering information about the distribution of fungi in The Netherlands. From its early days, the Dutch mycologist Frencken kept lists of records of field excursions. In the 1960s mycologists became aware of a change in mycota in The Netherlands (Bas, 1959, 1978), and assessment of this was one of the motives behind the conception of a national recording project in 1980. This project aimed at improving knowledge of the distribution and ecology of (macro)fungi. Because of organisational and financial obstacles it was not until 1984 that the project ran well. Since a recording project cannot work sufficiently without a checklist of the species to be recorded, the *Standaardlijst van Nederlandse macrofungi* was published in 1984 (Arnolds, 1984). This checklist contained an annotated list of the macrofungi to be expected in The Netherlands, and was based on literature, data from private and institutional herbaria and expert judgement. In relation to the publication of this preliminary Red Data list the Dutch Mycological Society organised a meeting in 1984 on the changes in the mycota of The Netherlands. At this meeting a great deal of scientific and statistical evidence for the decline of macrofungi was presented (Arnolds, 1985).

The recording project was a collaboration of the Dutch Mycological Society, the Institute of Forestry and Nature management, the Biological Station Wijster and the IKC Naturemanagement, and was financially supported by the Ministry of Agriculture, Nature Management and Fisheries. At present only the Dutch Mycological Society runs the project. The database now contains over a million records. Besides records from the recording project, volunteers have entered many data of older date into the database, derived from literature and herbaria. Many members of the Society, as well as naturalists from outside the Society, contribute to the database and field excursions are often planned in areas of The Netherlands which are under-recorded.

Recently, a monitoring project has been started, organised by the Dutch Mycological Society and the Central Bureau for Statistics, also financed by the Information and Knowledge Centre Naturemanagement (IKCN). This is aiming to monitor a limited number of rare and threatened species in woodlands on sandy soil (Arnolds & Veerkamp, 1999). This project is based on yearly counts in permanent plots of 1000 m^2 and hopefully will permit an evaluation of fungi in woodlands that is statistically more reliable than before.

Important publications on distribution of fungi

The mapping project has enabled publication of a new checklist (Arnolds, Kuyper & Noordeloos, 1995), which was written by 24 authors, including both amateur and professional mycologists. This new checklist contained data on nomenclature, distribution, rarity and ecology of the recorded macrofungi. Also, after publication of the preliminary Red Data list in 1989 (Arnolds, 1989), it has been possible to compile a new Red Data list of threatened fungi applying the official guidelines of the International Union for Conservation of Nature and natural resources (IUCN), by the direction of the Ministry of Agriculture, Nature Management and Fisheries (Arnolds & Kuyper, 1996; Arnolds & van Ommering, 1996). A digital version of the new checklist with addition of the most recent Red Data list data is included in BioBase (Anon., 1997), a CD-ROM with coded data on many groups of organisms.

Another important publication was the distribution atlas containing detailed information of a selected number of 375 taxa on distribution, ecology and phenology and their changes (Nauta & Vellinga, 1995). The data allowed statistical analysis and showed once more the serious decline in macrofungi in The Netherlands. The project was made possible by a grant from the Ministry of Agriculture, Nature Management and Fisheries which was awarded to the Dutch Mycological Society.

Conservation of fungi

The Committee for Fungi and Nature Conservation gives advice and answers questions concerning conservation of fungi. When Environmental Impact Studies are initiated for projects in which nature is affected, the Committee verifies whether data are available concerning fungi and tries to ensure that no important habitats are destroyed. Popularising fungi and emphasising the important role fungi play in nature is one of the ways to achieve a better understanding and protection for fungi. For this purpose a popular booklet of 20 pages was published in 1989 (Anon., 1989).

At present the most important project of the Committee for Fungi and Nature Conservation is the project 'Mycological Reserves', in which a better protection for mycologically valuable sites is one of the goals. The procedure and background of this project are discussed in Chapter 6. The major tool for conservation of fungi in The Netherlands is the Red list. By counting the number of species occurring in a site or using a weighting-factor according to the amount of threat experienced by a species, the mycological value of a site can be established. By means of giving specific management advice to people or councils dealing with management it aims to protect and improve the mycota.

In contrast to other countries (see, among others, Chapter 18) protection of fungi is solely tackled through protection of habitats and sites, since it is believed that fungi themselves can hardly be protected independently of their habitat. An exception is the case where larger woodland fungi grow on trees that are threatened with removal but, then, even this is habitat destruction. No biodiversity action plans for fungi exist in The Netherlands. Instead, protection of fungi is pursued by promoting understanding of fungi at all management levels and by indicating threatened habitats and sites.

Collecting fungi

In The Netherlands collecting fungi has also become an issue for discussion. In some nature reserves commercial picking is experienced as a problem (Anon., 2000). In principle, the Society wants to stimulate study of fungi but is not encouraging people to pick large quantities of mushrooms. Direct damaging effects to the mycelia by picking the fruit bodies has never been proven, but in an overcrowded country like The Netherlands the visual damage and the indirect damage resulting from trampling or disturbance of the forest floor is too great to encourage harvesting of fungi. The Society takes the position that fungi should only be collected in limited quantities when it is necessary for scientific, identification or educational purposes. In most nature reserves, strict rules already prevent the picking of mushrooms, and field excursions aiming at recording of fungi need special permission to collect the necessary specimens for identification.

References

Anon. (1989). *Paddestoelen.* Commissie Paddestoelen en Natuurbeheer, Nederlandse Mycologische Vereniging: Wijster.

Anon. (1997). *BioBase 1997, Register Biodiversiteit.* Centraal Bureau voor de Statistiek: Voorburg/Heerlen.

Anon. (2000). Grootschalige paddestoelenpluk, een toenemend probleem? *Coolia* **43**, 45–46.

Arnolds, E. (1984). Standaardlijst van Nederlandse macrofungi. *Supplement, Coolia* **26**.

Arnolds, E. (1985). Veranderingen in de paddestoelenflora (Mycoflora). *Wetenschappelijke Mededelingen van de Koninklijke Nederlandse Natuurhistorische Vereniging* **167**, 1–101.

Arnolds, E. (1989). A preliminary Red Data List of macrofungi in The Netherlands. *Persoonia* **14**, 77–125.

Arnolds, E., Kuyper, T. W. & Noordeloos, M. E. (1995). *Overzicht van de paddestoelen in Nederland.* Nederlandse Mycologische Vereniging: Wijster.

Arnolds, E. & Kuyper, T. W. (1996). *Bedreigde en kwetsbare paddestoelen in Nederland. Basisrapport met voorstel voor de rode lijst.* Nederlandse Mycologische Vereniging/Biologisch Station LUW: Wijster.

Arnolds, E. & van Ommering, G. (1996). *Bedreigde en kwetsbare paddestoelen in Nederland, Toelichting op de Rode lijst*. IKCN, Wageningen.
Arnolds, E. & Veerkamp, M. T. (1999). *Handleiding Paddestoelenmonitoring*. Nederlandse Mycologische Vereniging, Baarn.
Bas, C. (1959). Verdwijnende landschappen en hun fungi. *Coolia* **6**, 7–8.
Bas, C. (1978). Veranderingen in de paddestoelenflora. *Coolia* **21**, 98–104.
Nauta, M. M. & Vellinga, E. C. (1995). *Atlas van Nederlandse paddestoelen*. A. A. Balkema Uitgevers B.V.: Rotterdam.

22

Fungal conservation in the 21st century: optimism and pessimism for the future

DAVID MOORE, MARIJKE M. NAUTA, SHELLEY E. EVANS & MAURICE ROTHEROE

This book has examined a variety of problems associated with fungal conservation. We have tried to go beyond mere debate by including constructive guidance for management of nature in ways favourable to fungi. The geographical range of the examples presented is from Finland in the North to Kenya in the South, and from Washington State, USA, in the West to Fujian Province, China, in the East. Our authors suggest solutions that are equally wide ranging: from voluntary agreements, through 'fungus-favourable' land management techniques and on to primary legislation. Taken together, the book offers practical advice on how to include fungi in conservation projects in a range of circumstances.

The book has its origin in a British Mycological Society Symposium at the Royal Botanic Gardens in Kew, UK, in November 1999, combined with a selection of papers delivered at the XIIIth Congress of European Mycologists, held in Alcala de Henares, Spain, in September 1999. There were lively discussions at both meetings but verbal discussion is difficult to represent in writing. We felt it was important, however, to attempt to convey some feeling of the discussions that occurred during the meetings, so we asked contributors to respond, briefly, to the question 'Are you optimistic or pessimistic about fungal conservation in the 21st century?'

Our contributors replied from their different standpoints and their replies are given here without editorial interference from us. The replies are arranged in the same order as the authors' chapters in this book and the opinions expressed are those of the individual authors.

'Are you optimistic or pessimistic about fungal conservation in the 21st century?'

From Régis Courtecuisse, Lille, France

I can say that I am reasonably optimistic, in the sense that more and more people, especially those involved in nature management (foresters, nature managers, officials from government or nongovernment organisations) turn out to be interested and concerned with the role of fungi in ecosystems. Some examples of positive impact of fungal conservation (in The Netherlands for example) are encouraging to go further in this research and activity. Of course, this good impression on the future of fungal conservation is counterbalanced by the global threat to the environment, especially to the forests world-wide. These global problems (forest cutting as well as global pollution on the other hand) must be also considered and addressed, even if it is evidently more difficult to hope for a serious improvement in the situation from this point of view. Local (I mean at nonglobal scale) actions for fungal conservation and for the conservation of their natural habitats may be successful, especially in the perspective of future improved collaborations with nature management professionals of any kind. That is encouraging, but the global environmental problems will probably go further in degradation and threat. What makes me still optimistic, despite this globally pessimistic remark, is the actual possibility to improve the collective and federated actions and efforts, among mycologists but also among naturalists and nature managers. That's what I'll try to do and organise in the near future, as chair of the ECCF and the Specialist Group for Fungi (SSC/IUCN).

From Randy Molina, Corvallis, Oregon, USA

Overall, I am optimistic regarding fungal conservation, but mycologists have a lot of work to do in educating the public, government management agencies, and other biologists as to why and how. One pitfall to avoid is advocating 'single species management' because there are simply too many fungal species for this approach. Instead, our focus should be on maintaining or building habitat at the appropriate scale to provide for the development and normal function of fungal guilds and communities. To do so we must aim our field research at questions and scales relevant to conservation issues. These include successional development of fungal communities, disturbance thresholds and adaptations, habitat preferences, and ecological amplitude. Fundamental work is also needed at the population level to understand how we can incorporate genetic data into our concepts of fungal species viability. Mycologists must begin to think more broadly in scale and scope in regards to fungal conservation. We must become better versed in the quickly growing field of conservation biology, publish our work in international conservation journals, and look for opportunities to work with other conservation biologists and field ecologists in integrated studies with common goals. Mycology cannot stand alone in its pursuit of species conservation. Success will only come about when we demonstrate and communicate the great importance of maintaining fungal populations to the sustainability of our world's diverse ecosystems.

From Eef Arnolds, Beilen, The Netherlands

I am rather optimistic about the fungi in forest communities in Europe. There is an international trend to maintain remnants of old-growth forests and to create new strict reserves. The exploitation of forests tends to become less rigorous and more often based on ecological principles and durability. On the other hand, continued economic growth may have a negative impact on environmental quality that is harmful for many fungi by expansion of polluted areas. I am pessimistic about the future of fungi in other habitats. Intensification of agriculture will have a strong negative impact on the environmental quality and the existence of semi-natural habitats. Expansion of the EU will lead to great losses in the rural habitats in eastern Europe. These developments may be compensated by increased efforts for conservation of valuable habitats, but only in part. However, the greatest losses will take place in the tropics during an ever increasing process of habitat destruction, stimulated by the modern religion of the free market, invented in the rich countries. I have great concern about the future of field-mycology as a scientific discipline in Europe. Without experts we shall simply not know what happens to the mycota.

From Erast Parmasto, Tartu, Estonia

Very optimistic. Fungal species will be included in the European Red Data list soon – maybe some few species in the beginning, a few more later. Fungi will be mentioned in international conventions. It will take much more time (maybe, a dozen of years), but hopefully European mycologists will agree which species will be there in a European Red List (for 'internal use'); 25, 50 or 100 species will be selected then for distribution mapping. Actual or 'true' conservation is mainly possible in nature reserves. It depends on mycologists themselves: are they interested to take part in all-species inventories of protected areas? Do they care to spend some time for discussions with managers of reserves? Have they something to propose to them for better protection of habitats rich in fungi? Do they quarrel with ornitholocracy- or ornithologarchy-minded persons on equal rights of all endangered living beings to be protected? Do they dare to send articles on fungi to newspapers and popular scientific journals, to participate in TV shows in a studio or even in a forest stand? Are they ready to lobby MPs for selecting a State Mushroom (as we have State Flowers in many countries), using for this action of the legislature? Fate of fungal conservation is in our own hands.

From Leo Jalink & Marijke Nauta, National Herbarium of The Netherlands

We are moderately optimistic for the situation in The Netherlands. There is more and more attention for fungi in management plans now, and the general (changed) approach to nature conservation is usually beneficial for fungi as well. Of course, on the other hand there is an ongoing habitat destruction, and some habitats are still severely threatened (nutrient-poor grasslands, for example) but, at least, more attention is being paid to fungi.

From David Arora, Santa Cruz, California, USA

Since I didn't hear the discussion, and am about to leave for the airport, I will just have to leave it at, 'Optimistic, providing habitat is conserved . . . '.

From Siu Wai Chiu, Chinese University of Hong Kong, China

China is one of the megadiversity countries and is also one of the most highly densely populated countries by man. Another unique feature of China is its diverse geography and thus the fauna, flora and mycota are very rich. However, there is always a conflict between human interests and the protection of natural resources especially when the present status of China is still that of a developing and poor country. However, the Government plays an active role in policy making, practising and running the conservation plans. With the continual and expanding international effort in conserving the natural habitats in China, the international demand for natural products from sustainable industry, the open door policy in improving communication and education with the outside world, the openness in accepting and assimilating favourable culture by the Chinese, China is expected to actively participate in environmental protection. As the Chinese have the oldest history of mushroom cultivation, with over 30 fungal species being used in traditional herbal medicines, and the country is now the champion mushroom producer and exporter, the Chinese people will definitely guard, protect and explore the native fungal diversity. For fungal diversity in this biosphere, although habitat loss is still a major problem for species extinction, fungi being mainly saprotrophs, could survive in various artificial habitats. With the increasing interest and recognition of the potential for exploitation of fungi, and a global view emphasising sustainable development, humans will definitely make greater efforts to conserve biodiversity, including fungal biodiversity, into the twenty-first century.

From Maurice Rotheroe, Lampeter, South Wales

Like some other contributors to this chapter, I am ambivalent. The 'cons' seem to be just as compelling as the 'pros'. I share many of my colleagues' misgivings about the global future of fungal conservation, and any possible stemming of the tide of 'progress'. But if I confine my remarks to the British Isles, then the outlook is considerably more optimistic. Conservation is now established as a priority on the agenda of the British Mycological Society. Since 1996, the Society has had a Conservation Special Interest Committee. The Conservation SIC is now one of the two most active committees of the Society. Closer co-operation with CABI Bioscience and with the Mycological Herbarium of the Royal Botanic Gardens, Kew, has played a part in the establishment of a BMS database which is rapidly increasing in size and is in constant use as a resource. Similarly, the Society has given its backing to the ambitious Kew project to produce a new Checklist of British fungi. Both of these products are essential tools to the fungal conservationist. Concurrently with these developments, the Society has forged a strong link with the new National Botanic Garden of Wales. It has given £5000 of its Conservation Fund to the creation and curation of a mycological herbarium at the new garden and the funding of work which will eventually lead to publication of a Mycota Cymru (Fungus Flora of Wales). Discussions are proceeding on the setting up of a so-called 'Mycodome' on the NBG Wales site, to interpret and celebrate the concept 'Fungi Mould Our Lives'. The BMS is well represented on the steering committee of this unique £2 million international initiative. If it becomes a reality, fungal conservation will be a major element in its brief. When I became BMS Conservation Officer in 1996, my 'manifesto', was strongly based on the idea of communication – of greater collaboration and contact with other biological disciplines and with governmental and non-governmental environmental agencies, as well as between mycologists, within and outside the BMS. All of us appeared to be working towards a common goal but, with a few notable exceptions, there was little communication or co-operation between those involved, while routine information exchange was almost nonexistent. This situation has changed considerably in the past five years. Government agencies and NGOs are involved in a wide variety of different initiatives aimed at improving the representation of fungi in the biodiversity picture. There are now some 30 local recording groups well dispersed over the British Isles, who are not only recording and mapping but also liaising with local environmental agencies. In more recent times a new effective and extremely businesslike player has moved confidently into the fungus conservation arena. Plantlife, the wild-plant conservation society, set up Plantlife Link in 1992 in order to 'advance plant conservation by facilitating the exchange of information between groups involved in plant conservation and to provide the context for strategic planning and co-operative venture'. Fungi were regarded as 'honorary plants', so the forum included representatives from the BMS, the Association of British Fungus Groups and several individual mycologists. Because fungi began to feature more and more on agendas, an offshoot, the Fungus Conservation Forum was formed in 1999. This committee already has a large and active agenda and fulfils my 'manifesto' ambitions admirably because it puts fungal conservation at the heart of the thinking of so many different environmental agencies. I see Plantlife and its dynamic links as being a very important catalyst in advancing the cause of fungal conservation in Britain, certainly in the early years of the twenty-first century.

From Martin Allison, Bedfordshire, UK

Most fungal population declines and extinctions across Northern Europe are accounted for by problems deep-rooted in the fabric of modem society. The insatiable demand for fossil-fuelled energy produces unacceptable levels of atmospheric pollutants. The ever-increasing human population creates new housing schemes spilling onto green-field sites. Intensive farming regimes controlled by inappropriate agricultural policies cause massive declines of once-common farmland birds, insects, flowers and fungi. The logical conclusion might well be that the future looks dismal, for wildlife and for us. Is there room for optimism within such a bleak landscape? I believe there is. Coming from a background of bird conservation I have heard many success stories over the years. The osprey's return to the Scottish Highlands, the avocets to Minsmere, red kites back in England. Extensive tracts of heathland and reedbed restored, increasing populations of rare species such as woodlark, Dartford warbler, Cetti's warbler, bittern and bearded tit. All this has been achieved through dedicated and enthusiastic conservationists, backed by sound scientific research and increasingly more meaningful legislation. Fungal conservation has progressed in leaps and bounds over a very short period of time. There are now Red Data lists for most European countries, the most-threatened fungi are appearing within biodiversity action plans, and in the UK the habitat and survival requirements of rare fungi are being addressed by bodies such as Plantlife, English Nature, the RSPB and of course the British Mycological Society. To cap this, early in the New Year the UK Government announced a long-needed new Bill to secure better legal protection for wild plants and animals. The mere fact that this volume on Fungal Conservation exists gives some hope for the future. Through research, education, lobbying and that all-important 'in-the-field' enthusiasm many of the mycological goals should eventually be within reach.

From Giuseppe Venturella, Palermo, Italy

I am really optimistic about the possibility that the safeguarding actions for the mycota that need to be done in many European countries will be achieved in the twenty-first century. If this desirable goal is to be achieved, though, I think it is necessary to co-ordinate planning among the different Associations and Institutions working towards the conservation of fungi, such as BMS, ECCF, OPTIMA, CEMM, Planta Europa, Amateur Mycological Associations, etc.

From Heikki Kotiranta, Finnish Environment Institute, Finland

The intensive survey of forests done in Finland over the last ten years led to protection of large areas of old-growth forests. Old-growth forests in National Parks, nature reserves and some other areas are now protected by law and incorporated in nature conservation programmes. The new forest cutting recommendations, that always some amount of wood (often old, large or already dead trees) is left to the managed area, may help some species to survive even in commercially managed forests. However, most of the protected areas are situated in the northern part of the country, and the area protected in southern Finland is still too small, so further actions in southern Finland are required.

From Paul Cannon, CABI Bioscience, Surrey, UK

There are opportunities for both optimism and pessimism. On the positive side, conservation policy is changing dramatically from individual species-oriented approaches to ecosystem preservation, resulting in the conservation of many fungi by default. The value of fungi to the environment and ecological processes is being more widely appreciated, and the role of pathogenic species in the natural environment is being reassessed. The potential is bright for sensitive and efficient detection and monitoring of individual species during the next century, with ever more sophisticated molecular tools being developed. Pessimists are likely to be rewarded also. The dramatically increasing pressure on land, especially on natural habitats in the tropics, is probably causing fungal extinctions on a daily basis. Sustainable exploitation of natural resources will continue to be aspiration rather than reality as economic pressures force short-term decision-making. Global warming will pose added challenges for species survival. More specifically, funding for systematics shows little sign of reversing its downward trend, despite its critical role in identification and increased emphasis on the environment following the Rio conference. The continuing lack of interaction between fungal systematics experts and ecologists is also depressing, especially in areas such as sampling and monitoring procedures. Most fungal survey work is still carried out on an unscientific serendipitous basis, which makes comparison of results almost impossible. The key to ensuring a prominent role for fungi in conservation is to promote interdisciplinary links between scientists, and a holistic vision of fungi and their myriad roles in the ecosystem.

From Vincent Fleming, Peterborough, UK

We should be optimistic, for the UK anyhow. There is now increasing recognition that fungi are subject to the same pressures as, and are of equal worth to, other wildlife. This change in attitude can be attributed to the recent emphasis on biodiversity and it marks a radical shift from earlier assumptions that the conservation of fungi would, at best, have to depend on their chance representation in sites specially protected for other, more familiar, interests. The financial resources directed to fungi by conservation agencies have also increased, yet fungi still lag behind many other taxonomic groups. Realistically, they are likely to continue to do so: we have a shortage of specialists, a paucity of time-series data from which to estimate trends, and little knowledge of how to manipulate populations. To achieve maximum gain, we should seek to identify and highlight those circumstances where the needs of fungi differ from a 'normal' conservation approach. However, we should not expect to achieve total knowledge nor to delay action until we do so. We must use the best available information to guide and influence positive action and to stimulate public interest. For parallels and parity, we should look less to vascular plants and more to the conservation of invertebrates, with which fungi share many similarities: namely, the large numbers of taxa, difficulties of identification, and the transient appearance of observable life-stages. We have a considerable task ahead and optimism does not guarantee success, but to be pessimistic is not to try.

From Alison Dyke, Edinburgh, Scotland

From a Scottish perspective, and particularly on the subject of harvesting wild mushrooms, the future of fungal conservation seems fairly rosy. There is real interest in fungal conservation among all those involved in harvesting wild mushrooms – after all for some it is their livelihood and they do have a long term perspective. Effort is being made to survey important habitats and species groups, e.g. waxcap grasslands and tooth fungi. There are obviously some areas of concern, where important habitats are being lost, where the funds to do research are hard to come by, etc. It is these areas where it is important that positive action is taken and it is everyone's responsibility to ensure that this happens.

Index

Abies 19
abundance 37, 105
Acanthophysium 36
acid rain 226
action plans 201, 245
Aecidium 205
Agaricus 157, 158
Agenda 21 210
agriculture 182
airborne fungi 188
Albatrellus 36
Aleurodiscus 86
allergen 188
Alnus 19
Alpova 36, 97, 100
Amanita 157–160
amenities 21
Amylocorticium 86
anamorph 199, 205
ancient woodland 148
ant-nest symbionts 200
Aphyllophorales 177–181
Arbutus 19
Arcangeliella 36, 97, 100
Arctostaphylos 19
Armillaria 214, 215
artificial logs 117
Asterodon 86
Asterophora 36
Astraeus 159
Aureoboletus 159

Baeospora 36
Balsamia 36, 97, 98
Bankera 148, 149
Barssia 97–99, 102
basidiospore dispersal 117
Battarrea 214, 227
BIG (Biodiversity Information Group) 210
biodeterioration 188, 191
Biodiversity Action Plan 209–218
bioindicators 14
biological records 166, 167
biotechnology 191
biotrophic 201, 203, 204
birds of prey 24
BMS (British Mycological Society) 121, 217, 223, 251, 253
boletes 33
Boletopsis 214
Boletus 36, 47, 86, 108, 148, 149, 153, 154, 157–160, 214, 216, 217, 236
Bondarzewia 36, 76
bracket fungi 152
Bridgeoporus 36
British Council 168, 171
Bryoglossum 36
Buglossoporus 214, 217
Byssocorticium 151

C/N cycling 199
calcicolous grassland 121
calcifugous grassland 121
Calocybe 160
Calvatia 109
Cantharellula 150
Cantharellus 33, 36, 47, 107, 108, 159, 160, 236
Castinopsis 19
Catathelasma 36
CCW (Countryside Council for Wales) 120
cep 236
Cercospora 205
Ceriporia 86
Chalciporus 36
Chamaecyparis 40
Chamonixia 36, 97, 100
champions 210, 213
change in land use 70

chanterelle 50, 107, 236
charcoal 158
checklist 10, 67, 90, 167, 184, 194, 243
CHEG profile 125
China 106, 111–118
Chinese forests 112
Choiromyces 36, 97, 98, 158
Chromosera 36
Chroogomphus 36
Chrysomphalina 36
Clavaria 74, 121, 129, 138, 150
Clavariadelphus 36, 86, 159
clavarioid 120, 138–140
Clavicorona 36
Clavulina 36
Clavulinopsis 121, 129, 138, 139
climate 45, 128, 228
Clitocybe 36, 157
coastal dunes 136
coastal forest 19
code of conduct 219, 235, 238, 239
coelomycetes 204
Colletotrichum 202, 204
Collybia 36, 158
commercial harvesting 9, 21, 22, 26, 46–49, 52, 53, 105–110, 145, 153, 156, 158, 160, 219–222, 226, 231, 232, 236, 245
communities 25–28, 30, 68, 213
computer records 169
computing resources 164
conservation efforts 54
conservation policy 223–241
conservation value 123, 125, 127
contact point 210, 213
Convention on Biological Diversity 209
coppice 153
Coprinus 158
Cordyceps 36
Coriolopsis 159
Cortinarius 36, 148, 149, 151, 157, 159
Corylus 19, 149
Country Group 210
Crataegus 140
Craterellus 36, 108
cryptic species 21
Cryptomeliola 204
Cuba 182–196
Cudonia 36
cultivated strains 113, 114
cultivation method 111
culture collections 13, 160, 172, 183, 192, 198
CWD (coarse woody debris) 27
Cyphellostereum 36

Darwin Initiative 185, 192, 200, 202
dead wood 152
deforestation 112, 184
degradation 188, 191, 199
Dentipellis 86
Dermocybe 36
Dermoloma 129
Destuntzia 36
DFID (Department for International Development) 162
Dichostereum 36
Diospyros 205
Diplomitoporus 86
directory of scientists 169
distribution 67, 68, 72, 92, 147, 168, 200, 215, 225, 242–244, 249
disturbance 27, 99, 102, 227
diversity 22, 29, 30, 32, 35, 37, 69, 73, 77, 85, 109, 114, 118, 134, 142, 154, 156, 163, 184, 185, 187, 197, 199, 206, 209, 211, 213, 245, 251, 252
dune slacks 137
dunes 147
Dutch Mycological Society 89, 242–246

Earth Summit 209
earth-tongues 120, 136–143, 150
ECCF (European Council for the Conservation of Fungi) 7, 64, 68, 75, 95
ecological corridors 12
ecology 68, 99
ecosystem functions 23
ecosystem management 22
ectomycorrhiza 19, 23, 33, 45, 49, 51, 52, 69, 95, 145, 148, 152
education 118, 190, 240, 248, 253
Elaphomyces 36, 95, 97, 98
Elasmomyces 97, 100
endangered species 81, 128, 130, 201, 202
Endangered Species Act 20
Endogone 36
endophyte 187, 200–202
enoki 109
Enterographa 217
Entoloma 36, 74, 120–143, 149, 150, 153, 157–159
environmental quality 14
Eocronartium 86
ephemeral fruit bodies 42
epigeous fungus 33
epiphytes 201
Estonia 81
Eucalyptus 158
ex situ conservation 160, 198
extinct hypogeous species 102
exudates 23

Faerberia 158

fairy-club 120, 121
Fayodia 36
Fevansia 36
fieldwork 168
fire 27, 45, 48
fire suppression 29, 30
Fischerula 97, 98
Fomes 148
Fomitopsis 86
food webs 21, 24, 29, 200
foray 223, 224
forest disease 21
forest felling 8
forest fungi 19, 22, 23, 26
forest gaps 24
forest management 20, 25, 29
forest products 21
forest soil 200
Fremitomyces 204
Fujian Province 114, 115
fungal habitat 42
fungal records 187, 225
fungivores 24
fungus linkages 25

Galerina 36
Ganoderma 86, 159
Gastroboletus 36
Gastrosporium 97, 100
Gastrosuillus 36
Gautieria 36, 97, 100
Geastrum 86, 151, 159
Gelatinodiscus 36, 40
gene flow 31, 32
gene pool 114, 118
Genea 97, 98
genet 32, 116, 117
genetic drift 31
genetic homogeneity 114
genetic isolation 34
genetic variation 32, 33
Geoglossum 74, 138, 150
geographic scales 32
Geopora 97, 98, 157
global trade 106
Gloeophyllum 178
Glomerella 204
Glomus 36
Gomphus 36, 86
grassland 74, 120–143, 145, 147, 150, 215, 217, 228, 250
grassland soil 200
grassland types 121, 127, 129, 133
grazing 140–142, 153
grazing, cattle 136
grazing, rabbit 134
grazing, sheep 134, 136
green lists 172
greenhouse effect 8
grey lists 172
Grifola 109, 159
Guangdong Province 114, 115
guidelines 54, 93, 103, 133, 134, 141, 142, 181, 221, 229, 244
Guignardia 202
Gymnomyces 36
Gymnopilus 36
Gyromitra 36
Gyroporus 158

habitat 75, 86, 95, 99, 148
habitat action plans 210, 217
habitat conservation 11, 103, 184, 250
habitat destruction 8, 69, 249, 250
habitat indices 42
habitat loss 232
habitat management 38, 103
habitat manipulation 49
habitat modelling 42–45
habitat plans 211
habitat protection 244
habitat scale 248
HAP (Habitat Action Plan) 210
Hapalopilus 86
health audits 202
heartwood 152
heathland 150, 253
Hebeloma 36, 66, 157, 158
hedgehog mushroom 108
Helvella 36, 158
hen of the woods 109
Hericium 86, 149, 154, 159, 214, 217
heritage value 10
Hippophae 140
Histoplasma 189
historic sites 40
Hubei Province 114, 115
Hydnangium 97, 100
hydnelloid 216
Hydnellum 86, 149
Hydnobolites 97, 98
hydnoid species 149
Hydnotrya 36, 97, 98
Hydnum 36, 47, 108, 149
Hydrobotrya 98
Hydropus 36
Hygrocybe 66, 74, 120–143, 150, 153, 154, 159, 214, 215
Hygrophorus 36, 151, 153, 154, 158, 159
Hymenogaster 97, 100
hyphal length in nature 199, 200
hyphomycetes 204
Hypocreopsis 214
hypogeous fungi 24, 33, 44, 95–103, 227

Hypomyces 36, 108
Hysterangium 97, 100

Idaho 19
indicator species 14, 84–86, 127, 128, 132, 225
indicator-species profile 125
industrial mycology 191
information storage 166
Inocybe 157
interior forest 19
inventory 10, 35, 38, 42, 82, 83
ITE (Institute of Terrestrial Ecology) 152
IUCN specialist group 13
IUCN threat category criteria 11, 70, 72, 75, 177, 179, 203, 244

Jiangxi Province 114, 115

Kenya 197–208
key habitats 210
king bolete 108

labile information 165
Lactarius 108, 140, 148, 151, 157–159
land management 48, 54, 154, 228, 247
land use 145
land use change 9
land-owners 222
landscape scales 31, 34
Larix 19
laying yard 112
LBAP (Local Biodiversity Action Plan) 210
lead partner 210, 213
leaf spots 205
Leccinum 148, 157
legislation 8, 20, 21, 160, 173, 189, 191, 237, 239, 247, 253
Lentinula 111–118
Lepiota 151, 159
Lepista 157
Leptoporus 86
Leucangium 47, 109
Leucogaster 36, 97, 100
Leucopaxillus 158, 159
lichen 187, 199
lignin decomposition 199
limestone hills 182
Lindtneria 86
Lithocarpus 19
Lobelia 203
lobster mushroom 108
long list 210
Lucanus 217

machinery 153
Macowanites 36
Macrotyphula 158
management 21, 30, 31, 36, 47, 48, 93, 122, 142, 152, 153, 160, 226, 229, 247, 250
management guidelines 54, 93, 103, 133, 134, 141, 142, 181, 221, 229, 244
management history 124
management plans 89, 90
mangrove swamps 182
mapping 10, 67, 68, 168, 226, 243, 249
Marasmius 36
Martellia 36, 97, 100
matsutake 50, 106, 107
medical mycology 191
Melanogaster 97, 100
Melanoleuca 158
Meliola 204
mesotrophic grassland 121
microarthropods 24
microfungi 197–208
Microglossum 129, 131, 138, 150, 214, 215
middle list 210
migrations 8
mire 121
modelling habitat 42
molecular techniques 28
monitoring 35, 103
Morchella 47, 107
morel 50, 107
mowing 136, 141, 142, 153
Multiclavula 86
multispore spawn 117
mushroom 24, 29, 33, 47
Mushroom Code 221
Mutinus 159
mutualisms 21
mycelium 199
Mycena 36, 148, 150, 151, 157
Mycodome 251
mycological reserves 12, 90, 244
mycological sites, The Netherlands 92
mycophagy 33
mycorrhiza 25–28, 30, 73, 154, 187, 200, 202, 236
mycoses 188
mycotoxicoses 188
Myriostoma 158, 159
Mythicomyces 36

natural products 200
natural selection 31
nature management 136, 248
nature managers 93
nature reserves 89, 99, 144–155, 151, 181, 189, 198, 227, 245, 249

NBGW (National Botanic Garden of Wales) 123, 251
necrotrophic 201–203
Neolentinus 36
Neournula 36
Nivatogastrium 36
nomenclature 126
northern spotted owl 20, 29
Northwest Forest Plan 20, 21
nutrient cycling 21, 23
nutrient retention 23
NVC (National Vegetation Classification) 120, 121, 127, 152

Octavianina 36, 97, 100
old-growth forests 20, 22, 29, 35, 69, 73, 75, 81, 82, 84, 180, 181, 249, 253
old-growth grassland 74
old-growth woodland 92
Oligoporus 86
Ombrophila 148
Omphalina 36
Omphalotus 158, 159
optimism 247–255
Oregon 19
organic carbon sinks 23
Ossicaulus 158
Otidea 36
Oudemansiella 159
over-picking 226
Overseas Forays 224
Oxyporus 76
oyster mushroom 109

Pachyphloeus 97, 98
Panaeolus 140, 141
parasites 201
pasture 153
PCR–RFLP 28, 30
peasant-farmers 111
Perenniporia 86
pessimism 247–255
Peziza 141
Phaeocollybia 36
Phaeolus 86
pharmaceuticals 200
Phellinus 86
Phellodon 36, 86, 148, 149, 159
Phlebia 86
Pholiota 36, 158
Phomopsis 202
Physisporinus 86
Picea 19, 99
picking wild mushrooms 9, 21, 22, 26, 46–49, 52, 53, 105–110, 145, 153, 156, 158, 160, 219–222, 226, 231, 232, 236, 245
Picoa 97, 98
pine forest 147, 148, 153
Pinus 19
Pisolithus 159
Pithya 36
plant pathogens 24
plantations 150
Plectania 36
Pleurotus 158, 160
Pluteus 151
Podostroma 36
pollution 8, 9, 52, 69, 70, 145, 152, 184, 225, 228, 232, 252
Polyozellus 36
polysaccharide exudates 23
ponderosa pine 30
population 31–33, 68, 252, 255
population biology 22, 30, 113
population structure 116
Populus 19
Poronia 214
portabello 109
powdery mildews 205
primary threats 180
priority species 212
priority species or habitat 210
productivity 50, 51
Protea 203
protected areas 189
protection 103, 158, 160
Pseudaleuria 36
Pseudomerulius 86
Pseudotsuga 19
public awareness 190
puffball 109
Pulcherricium 159
Pycnoporellus 86

Quercus 19

raising yard 112
rakes 158
Ramaria 36, 159
Ramariopsis 129
ramet 32
random mating 34
rare species 30, 35, 38–43, 69, 81, 84, 85, 89, 91, 128, 130, 145, 158, 197, 200, 205, 224, 225, 228
recommendations 181, 229
red (rarity, endangerment and distribution) data lists 11, 35, 37, 64–67, 70–72, 74, 84, 121, 128, 130, 145, 146, 148, 158, 160, 172, 177, 227, 229, 231, 249, 253
red data lists, European 71, 96–103
red data list, Idaho 37
red data list, Estonia 86

red data list, Oregon 37
red data list, The Netherlands 90–92, 243, 244
reproduction strategies 69
rhizomorphs 23
Rhizopogon 33, 36, 97, 100
Rhodocybe 36
Rickenella 36
Rigidoporus 86
risk status 37
Rozites 148
RSPB (Royal Society for the Protection of Birds) 144–155
RSPB reserves (map) 147
rural development 106
rural economy 105
Russula 36, 108, 148, 151, 157

Salix 19
sampling 31, 41, 42
SAP (Species Action Plan) 210
Sarcodon 36, 86, 149
Sarcosoma 36, 86
Sarcosphaera 36
Schizophyllum 157
Sclerogaster 97
Scotland 219–222
second-hand computers 162
Sedecula 36
Senecio 203
Serpula 86, 178
Shaanxi Province 114, 115
shiang-gu 109, 111–118
shiitake 109, 111–118
short list 210
Sichuan Province 106
Sicily 156–161
silent protection 103
silviculture 29, 30, 51, 53
Sistotrema 86
site monitoring 103
Skeletocutis 86
small mammal 24, 29, 44
snowbanks 40
soil fungi 189
soil structure 23
somatic compatibility 117
sooty moulds 204
Sowerbyella 36, 86
Sparassis 36, 108
Spathularia 36
spatial scales 32, 34
spawn production 113
species action plans 211, 214
species concepts 66
species of conservation concern 210
species statement 210, 212
Sphaerosoma 97, 98
spore dispersal 33
SSC (Species Survival Commission) 7
SSSI (Site of Special Scientific Interest) 120
Stagnicola 36
Steccherinum 86
Stephanospora 97
Stephensia 97, 98
Strobilomyces 149, 151, 159
succession 26
Suillus 33
survey 21, 28, 36, 39–41, 120, 134, 203, 216, 253
sustainable management 26, 48, 50, 52, 109, 118, 219, 250
symbionts 27

taxonomists 10, 39
taxonomy 38, 66
teleomorph 199, 204
Tephrocybe 158
Thaxterogaster 36
The Netherlands 242–246
threat categories 11, 98, 100, 178
threatened habitat 72
threatened species 70, 96, 97, 128, 130,144, 145, 180
threats 184
Tibet 106
timber harvest 20, 27, 45, 47, 107
timber management 152
Tomentella 86
tourism 184, 194
traditional technology 111
training 39
Trametes 179
trampling 227
Tremiscus 36, 86
Trichoglossum 138, 150
Tricholoma 36, 47, 107, 148, 157–159
Tricholomopsis 36
tropical expeditions 224
truffle 24, 29, 35, 95, 103, 107
Truffle Hunt 224
Tsuga 19
Tuber 36, 47, 97, 98, 109
Tulostoma 159, 214, 215
Tylopilus 36

UK BG (UK Biodiversity Group) 210
UK Darwin Initiative 163, 167, 168, 170
Ukraine 162–176
UNESCO biosphere reserve 189
United Kingdom 209–218
USDA (US Department of Agriculture) 20, 22

USDI (US Department of the Interior) 20

voluntary agreements 247
voluntary code 219
Volvariella 151

Wakefieldia 97
Wales 120–135
Washington 19
waxcaps 120–143, 150, 215, 217
wild mushroom harvesting, issues 220
wild mushrooms 9, 21, 22, 26, 46–49, 52, 53, 105–110, 145, 153, 156, 158, 160, 219–222, 226, 231, 232, 236, 245
woodlands 147
woody debris 27, 30, 35, 45, 83

Xylaria 86
Xylobolus 86

Yunnan Province 106, 114, 115